Practical Clinical Biochemistry

Practical Clinical Biochemistry

Shruti Mohanty MD

Professor and Head
Department of Biochemistry
Kamineni Institute of Medical Sciences
Dist. Naretpally (A.P), India

Aparna Verma MD

Professor and Head
Department of Biochemistry
Mediciti Institute of Medical Sciences
Ghanpur, Medchal Mandal,
Dist. Rangareddy (A.P), India

JAYPEE BROTHERS MEDICAL PUBLISHERS (P) LTD

New Delhi • London • Philadelphia • Panama

Headquarters

Jaypee Brothers Medical Publishers (P) Ltd
4838/24, Ansari Road, Daryaganj
New Delhi 110 002, India
Phone: +91-11-43574357
Fax: +91-11-43574314
Email: jaypee@jaypeebrothers.com

Overseas Offices

J.P. Medical Ltd
83 Victoria Street, London
SW1H 0HW (UK)
Phone: +44-2031708910
Fax: +02-03-0086180
Email: info@jpmedpub.com

Jaypee-Highlights Medical Publishers Inc.
City of Knowledge, Bld. 237, Clayton
Panama City, Panama
Phone: + 507-301-0496
Fax: + 507-301-0499
Email: cservice@jphmedical.com

Jaypee Brothers Medical Publishers Ltd
The Bourse
111 South Independence Mall East
Suite 835, Philadelphia, PA 19106, USA
Phone: + 267-519-9789
Email: joe.rusko@jaypeebrothers.com

Jaypee Brothers Medical Publishers (P) Ltd
17/1-B Babar Road, Block-B, Shaymali
Mohammadpur, Dhaka-1207
Bangladesh
Mobile: +08801912003485
Email: jaypeedhaka@gmail.com

Jaypee Brothers Medical Publishers (P) Ltd
Shorakhute, Kathmandu
Nepal
Phone: +00977-9841528578
Email: jaypee.nepal@gmail.com

Website: www.jaypeebrothers.com
Website: www.jaypeedigital.com

Inquiries for bulk sales may be solicited at: jaypee@jaypeebrothers.com

Practical Clinical Biochemistry

First Edition: **2013**

ISBN 978-93-5090-465-7

Printed at : S. Narayan & Sons

To

The Medical Students

Preface

This book *Practical Clinical Biochemistry* is an endeavor to correlate Biochemistry with clinical cases and bridge the gap between basic science and clinical medicine. In an era of automation, manual tests have retained their distinct entity by virtue of being cost-effective and easy to perform procedures, thereby giving crucial information about a particular disorder. The undergraduates shall perform the tests and know the essence of doing so by understanding the principle behind the test and also its relevance in medical science.

This book has many illustrations going by the adage "a picture is worth a thousand words" to make the subject easy to understand and therefore retain longer. The summary tables are like the ready reckoner, recapping all the important test findings and implications, thereby rendering a student friendly approach.

A note on "point of care testing" (POCT) and "analytical errors" has been included to emphasize the role of simple bedside investigations and the importance of excluding/rejecting samples liable for giving erroneous results respectively. Techniques like Electrophoresis and Chromatography are the commonly applied procedures for diagnosis of various common liver and renal disorders, paraproteinemias, etc. by the former and detection of aminoacidurias by the latter.

We hope this book will contribute in integrating biochemistry with other clinical disciplines.

Shruti Mohanty
Aparna Verma

Acknowledgments

The authors would like to express their gratitude to all those who have directly or indirectly supported in completing this book. We are highly indebted to Prof. Pragna Rao, Manipal University for her constant encouragement and unabated boost without which it was difficult to initiate writing this book. We want to extend sincere thanks to all our colleagues for their valuable suggestions.

On a personal note, we would like to thank our families profusely whose blessings and love has enabled us to complete this book.

Last but not the least, we thank the Jaypee Brothers Medical Publishers (P) Ltd staff for translating our dream into reality.

Contents

Basic Laboratory Principles and Safety Measures

VOLUMETRIC EQUIPMENT

Most clinical chemistry procedures require accurate measurement of volume. Volumetric equipment should be used with solutions equilibrated at room temperature. For accurate work with pipettes and burettes, they should be rinsed out and drained well with some of the liquid to be measured. The filling of pipettes with poisonous, corrosive or volatile liquids should be done using a rubber teat.

Pipettes

In general, two main types of pipettes are used. The volumetric or transfer pipette is designed to deliver a fixed volume of liquid. They are calibrated to deliver the volume specified. The most commonly used sizes are 1, 2, 5, 10, 25 ml.

All these pipettes are calibrated **To deliver (TD)** a specific volume. **To contain (TC)** pipette are used for volumes smaller than 0.5 ml. The difference between TC and TD types of pipettes is that the former must be rinsed out after complete delivery in order to wash all the fluid contained in the pipette into the diluent.

The second type of pipette is the graduated or measuring pipette. These are of two types—one is calibrated between two marks on the stem (Mohr) and the other has graduation marks down to tip (Serological). They are not suitable for accurate measurement of volumes less than 2 ml. They are principally used for measurement of reagents.

Volumetric Flasks

They are primarily used in preparing solutions of concentrate. They are found usually in the following sizes: 5, 25, 50, 100, 250 ml, etc. Volumetric equipment should be used with solution equilibrated at room temperature.

UNITS AND REFERENCE VALUES

System of measurements used in clinical laboratories is the metric system. An international system of units was introduced in 1971 called SIU (Systeme International'd' Unites) which accepts metric system.

Mole (Molecular weight in grams) is the weight measurement in chemical language. Concentrations of solutions and reagents prepared are expressed as:

i. Percent solution: 1 gram of a solute in 100 ml of solution is 1 percent. Expressed as %, g/dl or g/100 ml.

ii. Molar solution: 1 mole (gram molecular weight) of solute in 1000 ml of solution is equivalent to 1 molar solution expressed in different ways as I M, 1 mol/L or 1 mole/L.

Biochemical tests results are expressed as:

i. For metabolites like glucose, urea, etc. **mg/dl or mmol/L.**

ii. Electrolytes (sodium, potassium) as **mmol/L or mEq/L.**

iii. Enzymes as **Units/L.**

One unit of enzyme activity is defined as 1 micromole of substrate utilized or 1 micromole of product formed in one-minute time at controlled conditions of temperature, pH and substrate concentrations.

Enzymes are sometimes expressed in conventional units e.g., Amylase as Somogyi units and Phosphatases as King Armstrong units.

REGULATIONS AND SAFETY PRECAUTIONS IN CLINICAL LABORATORY

Careless handling of apparatus and reagents is a common cause of laboratory accidents, resulting in burns or fires. To avoid these, a few basic rules must be followed.

1. Poisonous reagents, concentrated acids and alkalies must never be pipetted by mouth.
2. Dispensers or pipette pumps must be used.
3. Organic solvents should not be boiled directly on flame.
4. Reagents must be replaced in proper place after use. Stoppers and pipettes should not be mixed, to avoid contamination.

Any personal injury or swallowing of reagents must be promptly reported. For burns, affected part must be put in cold water and then apply petroleum jelly, olive oil or burn ointments and cover with dry gauze. Students must wear laboratory overalls for the practical class. They should maintain discipline and follow instructions carefully. Cleanliness is essential in all biochemical work. Glassware and reagents must be properly handled. Any breakage of glassware must be reported.

Students must bring an observation note book to record data. They are required to maintain a laboratory record note book in which experiment has to be recorded and signed by the teacher concerned every following week.

COLLECTION AND HANDLING OF SPECIMENS

While ordering for a particular investigation, the doctor must fill in the form legibly giving relevant clinical information. After drawing blood of the patient from a suitable vein with as little stasis as possible the correct container should be used to transport it to the laboratory for analysis.

COLLECTION OF BLOOD

In case of analysis of constituents in whole blood or plasma, the container usually contains an anticoagulant. Potassium oxalate at a conc. of about 1 to 2 mg/ml of blood is most widely used. Other anticoagulants are EDTA, heparin. But in the case of blood glucose estimation, in addition to oxalate, sodium fluoride at a conc. of 2 mg/ml of blood is employed. This inhibits glycolysis which otherwise would continue in the RBCs. For blood urea estimation fluoride should not be used.

In case the analysis has to be performed on a specimen of serum, the blood is transferred to a clean dry tube after the needle has been removed. The blood is allowed to clot for at least 10–15 minutes at room temperature. After clotting, the tube is centrifuged and the supernatant serum removed.

Hemolysis should be avoided as it interferes with chemical procedure; erythrocytes contain very different concentration of many substances from plasma (e.g. potassium). If hemolysis occurs, these substances leak out and false results will be obtained on plasma. Plasma or serum from hemolyzed blood is red and can be detected.

COLLECTION OF URINE

An accurately timed 24 hrs collection is essential. A preservative to prevent bacterial and fungal growth as well as one depending on the substance being estimated are added to the container.

After collecting the specimen in the proper container, correct labeling is essential. In case of true emergency, the laboratory should be appropriately informed. Hence, the need for proper preservation and quick handling of specimen.

SUMMARY CHART FOR THE COLLECTION AND DISPATCH OF SPECIMEN FOR BIOCHEMICAL TESTS

Tests	*Specimen collection*	*Stability*
Alanine Aminotransferase (SGPT)	3–5 ml clotted blood in dry glass container. Hemolysis interferes with test.	Stability in whole blood at RT for 3 hrs and at 2–8°C up to 36 hrs.
Albumin	2 ml clotted blood. Hemolysis interferes with test.	Stable in whole blood at RT up to 8 hrs in serum at 2–8°C up to 4 days
Alkaline Phosphatase	3–5 ml clotted blood. Hemolysis Interferes with test.	Stable in whole blood up to 12 hrs and in serum at 2–8°C up to 48 hrs.
Acid Phosphatase	3–5 ml clotted blood. Hemolyzed blood must not be used.	Poor stability in whole blood. Serum must be separated as early as possible. Assay within 1 hr or store frozen
Amylase	3–5 ml clotted blood.	Good stability in whole blood. Stable in serum at 2–8°C for 3 days

Contd...

Contd...

Tests	***Specimen collection***	***Stability***
Aspartate Aminotransferase (SGOT)	3–5 ml clotted blood. Hemolysis interferes with test.	As for alkaline phosphatase.
Bilirubin (Total)	3–5 ml clotted blood. Hemolysis interferes with test. Infants: 1–2 ml anticoagulated blood. (Lithium-heparin or EDTA).	Protect from light, stable in whole blood for 3 hrs and in serum or plasma at 2–8°C up to 12 hrs.
Calcium	1–2 ml clotted blood. Collect avoiding venous stasis. Patient must not be on EDTA therapy.	Stable in whole blood for 3 hrs and in serum at 2–8°C up to 72 hrs.
Cholesterol	3–5 ml clotted blood. Fasting sample must be collected.	Stable in whole blood up to 12 hrs and in serum at 2–8°C up to 72 hrs
Carbon dioxide (Bicarbonate), Blood pH, PCO_2	5 ml anticoagulated blood (lithium heparin) Avoid introducing air bubbles into sample.	Very poor stability. Plasma Must be separated and analyzed as soon as possible.
Creatinine	2–3 ml clotted blood. Hemolysis interferes with test.	Stable in whole blood up to 12 hrs. In serum at 2–8°C up to 24 hrs.
Electrolytes Sodium, Potassium Chloride	2–3 ml clotted blood or anti coagulated blood (Lithium-heparin). Hemolyzed blood not to be used. Do not collect blood from an arm receiving IV infusion.	Stable in whole blood at RT up to 1 hr. Do not refrigerate sample before removing serum.
Glucose	2 ml anticoagulated (Fluoride - Oxalate) blood fasting, post-prandial or random. Do not collect blood from an arm receiving IV infusion.	Stable in whole blood at RT up to 1 hour. Do not refrigerate sample before removing serum. Stable in serum/plasma at 2–8°C up to 24 hrs.
Inorganic Phosphorus	3–5 ml clotted blood hemolyzed blood must not be used.	Stable in whole blood at RT up to 3 hrs. In serum at 2–8°C up to 5 days.
Protein (Total)	3 ml clotted blood. Avoid venous stasis. Hemolysis interferes with test.	Stable in whole blood at RT up to 8 hrs. In serum at 2–8°C up to 4 days.
Uric acid	5 ml clotted blood.	Protect from day light stable in whole blood up to 12 hrs and in serum at 2–8°C for 72 hrs.
Urea	1–2 ml clotted blood. Plasma from EDTA or fluoride/oxalate blood can also be used.	Stable in whole blood up to 12 hrs. and in serum at 2–8°C for 48 hrs.

Abbreviation:
RT: Room Temperature

Chapter 2 Qualitative Analysis

CARBOHYDRATES

INTRODUCTION

Carbohydrates are compounds containing C, H and O. General formula for carbohydrate is $(CH_2O)_n$. All have C=O or CHO as functional group. Carbohydrates are defined as polyhydroxy aldehyde or ketone or substances which give these on hydrolysis.

They are classified as–

1. Monosaccharides (Glucose and Fructose),
2. Disaccharides (Lactose, Maltose and Sucrose), and
3. Polysaccharides (Starch, Glycogen)

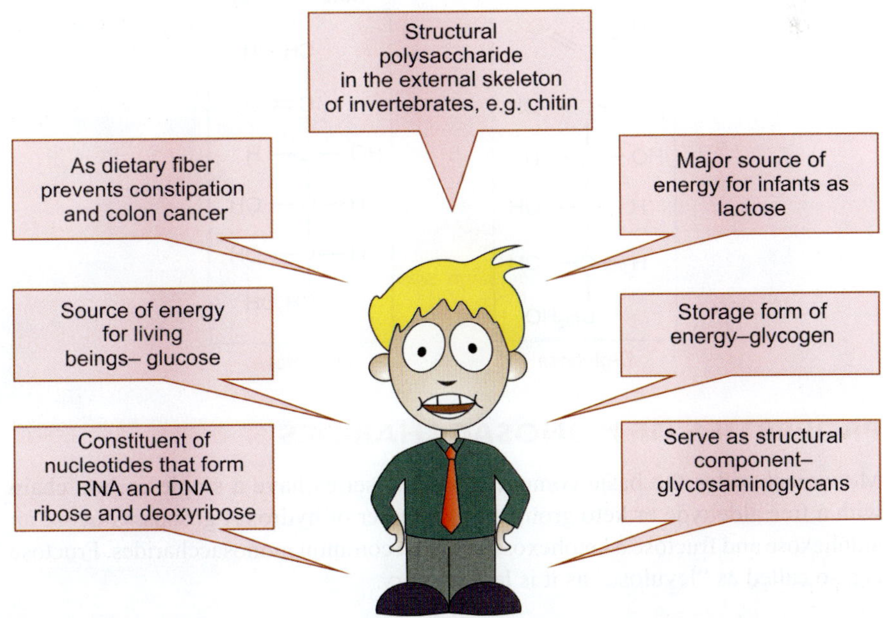

Carbohydrates are widely distributed in plants and animals; they have important structural and metabolic roles. Glucose is the major metabolic fuel of mammals (except ruminants) and a universal fuel of the fetus. It is the precursor for synthesis of all the other carbohydrates in the body, including **glycogen** for storage; **ribose** and **deoxyribose** in nucleic acids; and **galactose** in lactose of milk, in glycolipids, and in combination with protein in glycoproteins and proteoglycans. Diseases associated with carbohydrate metabolism includes **diabetes mellitus, galactosemia, glycogen storage diseases,** and **lactose intolerance.**

Types of carbohydrates

Classification is based on number of sugar units of total chain

Monosaccharides	single sugar unit
Disaccharides	two sugar units
Oligosaccharides	3 to 10 sugar units
Polysaccharides	more than 10 units

Chaining relies on "bridging" of oxygen atoms–glycosidic bonds.

Monosaccharide classification based on functional group

Aldose (H-C=O)

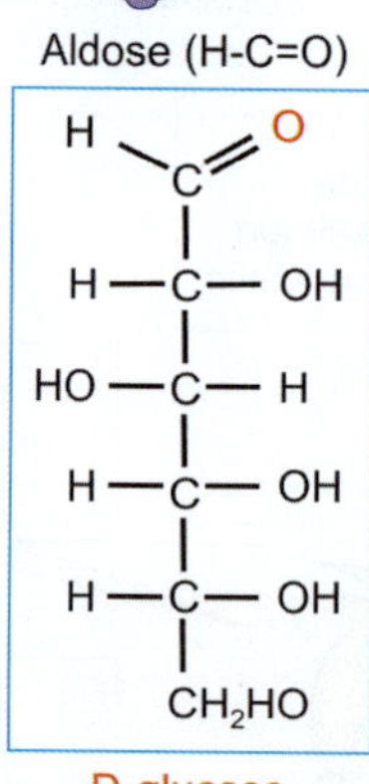

D-glucose

Ketose (C=O)

CH_2OH
C=O
HO—C—H
H—C—OH
H—C—OH
CH_2OH

D-fructose

PROPERTIES OF MONOSACCHARIDES

Monosaccharides the basic compound of this series have a single carbon chain with a free aldehyde or keto group and a number of hydroxyl groups. Glucose an aldohexose and fructose a ketohexose are most common monosaccharides. Fructose is also called as "levulose" as it is levorotatory.

Glucose and Fructose

D-glucose

- Glucose is an aldohexose sugar
- Common names include dextrose, grape sugar, blood sugar.
- Most abundant organic compound found in nature.

```
H
  \
   C=O
   |
H—C—OH
   |
HO—C—H
   |
H—C—OH
   |
H—C—OH
   |
  CH2OH
```

D-glucose

D-Fructose

- Fructose is a ketohexose
- Sweetest of all sugars

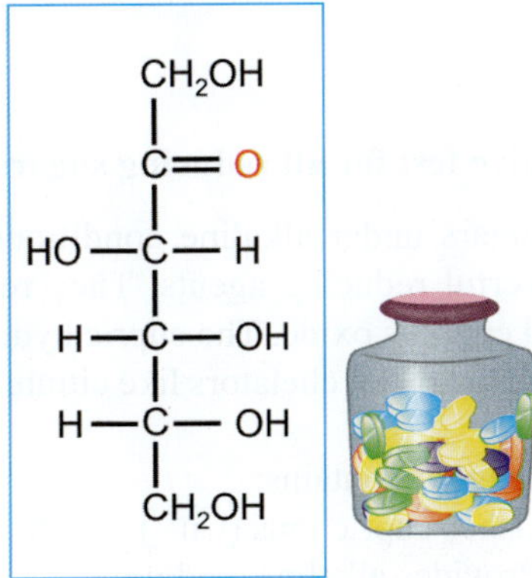

D-fructose

1. Molisch's Test

Reagent

1. Molisch's reagent (alpha-naphthol in 95% ethanol)
2. Concentrated sulfuric acid.

Principle: This reaction is a general test for all carbohydrates. Carbohydrates when treated with concentrated H_2SO_4 undergo dehydration to give furfural derivatives. These compounds condense with α-naphthol to form reddish violet colored products.

The oligosaccharides and polysaccharides are first hydrolyzed to the constituent monosaccharides which are then dehydrated. Pentoses yield furfurals and hexoses yield 5-hydroxymethylfurfurals.

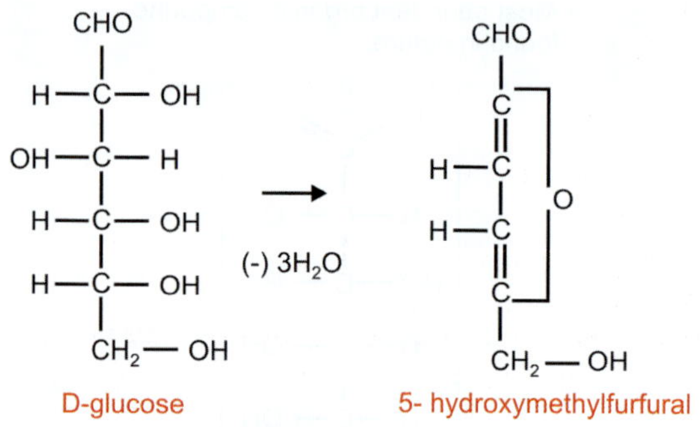

```
        CHO                                  CHO
         |                                    |
    H—C—OH                                     C——┐
         |                                    ||   |
   OH—C—H                               H—C   |
         |             ——→                    |    O
    H—C—OH                              H—C   |
         |          (-) 3H₂O                  ||   |
    H—C—OH                                     C——┘
         |                                    |
        CH₂—OH                               CH₂—OH
```

D-glucose   5- hydroxymethylfurfural

Formation of Furfural Derivatives

Procedure: To 2 ml of carbohydrate solution, add 2 drops of 1% alcoholic alpha-naphthol solution and mix. To this, add 2 ml of concentrated sulfuric acid along the side of the test tube to form a separate layer. A purple ring is formed at the junction of two layers.

The test is said to be positive only if a purple violet ring is formed at the junction.

Note: This test is also given by aldehydes and by formic, lactic, oxalic, citric acids and certain other organic acids.

2. Benedict's Test

Benedict's test is a sensitive **test for all reducing sugars.**

Principle:- Reducing sugars under alkaline conditions tautomerize and form enediols which are powerful reducing agents. They reduce the cupric ions of Benedict's reagent to red cuprous oxide. The cupric hydroxide formed during the reaction is kept in solution by metal chelators like citrate (or tartarate in Fehling's solution).

Benedict's qualitative reagent contains:

1. **Copper sulfate**: Furnishes cupric ions (Cu^{++}) in solution.
2. **Sodium carbonate**: Provides alkaline medium.
3. **Sodium citrate**: Prevents the precipitation of cupric ions as cupric hydroxide by forming a loosely bound cupric-sodium citrate complex which on dissociation gives a continuous supply of cupric ions.

Benedict's test is a semiquantitative test.

Glucose		1, 2 Enediol form		Fructose
H—C—O		H—C—OH		CH_2OH
H—C—OH		‖ C—OH		C=O
HO—C—H	⟷	HO—C—H	⟷	HO—C—H
H—C—OH		H—C—OH		H—C—OH
H—C—OH		H—C—OH		H—C—OH
CH_2—OH		CH_2—OH		CH_2—OH

Procedure: Take 5 ml of Benedict's qualitative reagent and boil for one minute in a test tube. No change in color or precipitation occurs. Now add 0.5 ml (8 drops) of carbohydrate solution and boil again for two minutes. Marked reduction is seen as yellow or red precipitate.

This test is employed as a routine test for examination of urine for sugar. Reducing sugar is found in urine of Diabetes mellitus, fructosuria, pentosuria and lactosuria.

(Refer Benedict's test under pathological urine)

Benedict's test- Principle

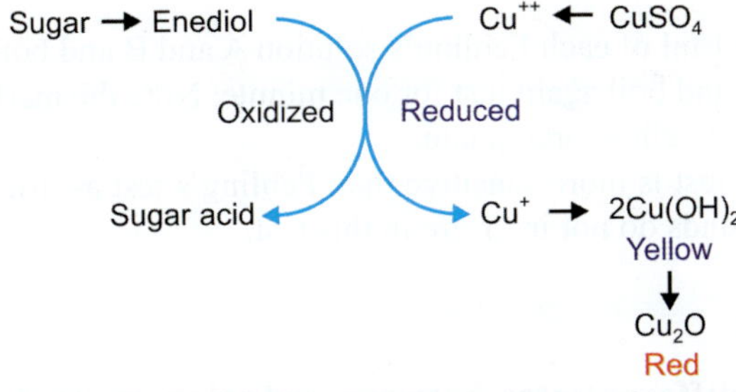

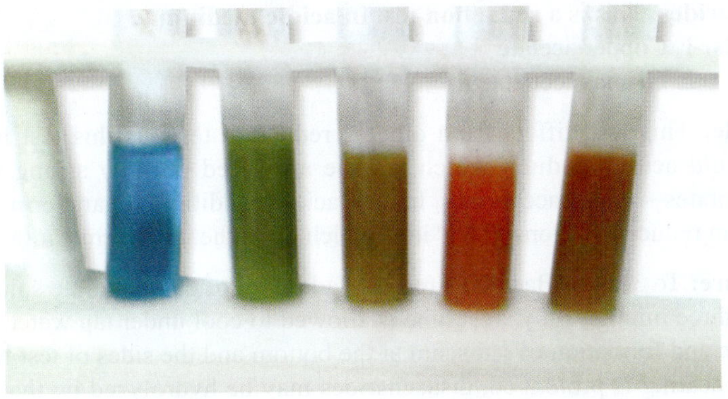

Color of the precipitate	*Approximate concentration of sugar in urine (gm %)*
Green (+)	0.1–0.5%
Yellow (+ +)	0.5–1.0%
Orange (+ + +)	1.0–2.0%
Red (+ + + +)	Above 2%

3. Fehling's Test

This is another **reduction test** to detect the presence of reducing sugars. Fehling's reagent differs from Benedict's reagent in that Fehling's reagent contains sodium potassium tartrate (Rochelle's salt) in place of sodium citrate.

Reagents:
Fehling's Solution Consists of:
Fehling's solution A: - Contains copper sulfate solution.
Fehling's solution B: - Contains potassium hydroxide and sodium potassium tartrate.
Prepare fresh by mixing equal volumes of Fehling's solutions A and B.

Principle: Carbohydrates with free aldehyde or ketone groups reduce copper sulfate to cuprous oxide forming yellow or brownish red colored precipitate.

Sodium potassium tartrate prevents precipitation of cupric hydroxide by forming a complex with the cupric ions. These complexes dissociate to provide a continuous supply of cupric ions for oxidation.

Procedure: Add 1 ml of each Fehling's solution A and B and boil. Now add 2 ml of sugar solution and boil again just for one minute. Note the marked reduction as indicated by red or yellow precipitate.
Note: Benedict's test is more sensitive than Fehling's test as uric acid, creatinine and other compounds do not interfere in this test.

4. Barfoed's Test

Barfoed's test **differentiates between reducing monosaccharides and disaccharides**. This is a **reduction test in acidic medium.**
Reagents: 1. Copper acetate
2. Glacial acetic acid.

Principle: This test differs from other 2 reduction tests as this test is carried out in mild acidic medium which will be answered only by strong reducing carbohydrates—monosaccharides. In mild acidic condition, sugars form enediols, which can reduce Cu^{+2} ions to Cu^{+} ions which up on heating form Cu_2O.

Procedure: To 3 ml of Barfoed's reagent, add 2 ml of carbohydrate solution and boil for three minutes only. Test tube is allowed to cool under tap water. There is reduction and formation of red scum at the bottom and the sides of test tube,
Note: If heating is prolonged, disaccharides may be hydrolyzed by the acid and the resulting monosaccharides will give the test positive.

5. Reduction of Methylene Blue

Methylene blue is used as an internal indicator in Lane and Eynon's volumetric method for determining reducing sugars.

Principle: Carbohydrates containing free aldehyde or ketone groups can reduce methylene blue to leucomethylene blue with the disappearance of the blue color.

The dye methylene blue exists in two forms. The oxidized form is colored blue, while the reduced form is colorless (Leuco – methylene blue). The color may re appear if solution is allowed to cool with shaking. This is due to re-oxidation by atmospheric oxygen.

Procedure: To 2 ml water, add one drop of 1% methylene blue and 5 drops of 5% sodium hydroxide and boil. The solution remains blue. Now add three drops of carbohydrate solution and boil again. The color disappears due to reduction.

6. Foulger's Test

Foulger's Reagent: Urea, Stannous Chloride, 40% H_2SO_4

Principle: This is a specific **test for ketohexoses.** Conc. H_2SO_4 dehydrates ketohexoses to form furfural derivatives which condense with urea in the presence of stannous chloride to give a deep blue color.

Procedure: To 3 ml of Foulger's reagent, add 0.5 ml of carbohydrate solution. Boil it vigorously for one minute and allow it to cool. A colorless solution is seen in case of glucose. Fructose gives a deep blue color solution.
Note: Prolonged heating converts aldoses to ketoses which yield deep blue color (false positive).

7. Seliwanoff's Test

Principle: This is also a specific **test for ketohexoses.** Concentrated HCl dehydrates ketohexoses to form furfural derivatives which condense with resorcinol to give cherry red color solution.

Procedure: To 3 ml of Seliwanoff's reagent, add 0.5 ml of carbohydrate solution. Boil for one minute and cool. A pink or red color develops for ketohexoses.
Note: Prolonged heating converts aldoses to ketoses which yield cherry red color (false positive).

8. Rapid Furfural Test

Principle: Ketohexoses are rapidly converted to hydroxymethyl-furfural by HCl and form purple colored complex with 1% alcoholic alpha naphthol. This is a **test for ketohexoses**.

Procedure: To 2 ml of glucose solution, add 2 drops of 1% alcoholic alpha-naphthol and 3 ml concentrated HCl, boil on the flame and cool. There is no color in case of glucose solution, however prolonged boiling will reduce the sugar.

In case of fructose a deep purple color is produced from the moment the solution starts boiling. The appearance of the purple color is of importance in the identification of fructose from glucose.

9. Bial's Test (Tollen's Orcinol Test)

Specific **test for pentoses** such a Ribose, Xylose
Bial's reagent:
Dilute solution of orcinol in 30% HCl

Principle: Conc. HCl acts on pentoses to form furfurals. This condenses with orcinol to give green color compounds.

Procedure: To 5 ml of Bial's reagent in a test tube 0.5 ml of carbohydrate solution is added and mixed. Solution is heated until it begins to boil. Green color solution is observed.

10. Osazone Test

Reducing sugars form characteristic crystals on reacting with phenylhydrazine in acidic medium.

Principle: Phenylhydrazine in acetic acid when boiled with reducing sugars form phenylhydrazones followed by the formation of osazones. The first two carbons (C1 and C2) are involved in this reaction. The sugars that differ in their configuration on these two carbon atoms give the same type of osazones, since the difference is masked by binding with phenylhydrazine. Thus, glucose, fructose and mannose give the same type of osazones, i.e. needle shaped.

Procedure: Dissolve about 100 mg of phenylhydrazine hydrochloride and 200 mg of sodium acetate in 1 ml of glacial acetic acid and 5 ml of sugar solution and heat in a boiling water bath for 30 minutes. Put off the flame and allow the test tubes to cool in the water bath itself. The cooling process should not be hastened by removal from water bath or adding cold water to it.

As the solution cools, crystals of glucosazones or fructosazones separate out as a rich crop of yellow crystals. Mount a few crystals on slide and cover with a cover slip and examine under the microscope. (The cover slip should be placed gently on the crystals without applying any pressure, otherwise the crystals will be broken and lose the characteristic shape). **Glucosazones** and **Fructosazones** both appear as **needle-shaped** crystals.

Needle-shaped crystals arranged like a broom
Glucosazone

Sunflower-shaped or petal-shaped crystals of **Maltosazone**

Hedgehog or "pincushion with pins" or flower of "touch- me- not- plant" **Lactosazone**

Figs 2.1 A to C:

Shapes of Osazones Under Microscope

$$\underset{\text{D-glucose}}{\begin{array}{c} CHO \\ | \\ H-C-OH \\ | \\ (CHOH)_4 \\ | \\ CH_2OH \end{array}} \xrightarrow[HCl]{C_6H_5NHNH_2} \underset{\text{D-glucose phenyl hydrazone}}{\begin{array}{c} CH=NNHC_6H_5 \\ | \\ H-C-OH \\ | \\ (CHOH)_4 \\ | \\ CH_2OH \end{array}}$$

$$\downarrow 2\ C_6H_5NHNH_2$$

$$NH_3 + C_6H_5NH_2 \qquad \underset{\text{D-Glucosazone}}{\begin{array}{c} CH=NNHC_6H_5 \\ | \\ C=NNHC_6H_5 \\ | \\ (CHOH)_4 \\ | \\ CH_2OH \end{array}}$$

Glucose and Fructose

- Both answer Molisch test
- Benedict's test, Fehling's test, methylene blue test are positive
- Barfoed's test is also positive
- Seliwanoff's test and Foulger's test are negative for glucose (aldo-sugar) and positive for fructose (keto-sugar)
- Both glucosazones and fructosazones are needle-shaped crystals.

PROPERTIES OF DISACCHARIDES

Two monosaccharides can combine by a glycosidic linkage to form disaccharides. The union can take place by fusion of a free aldehyde or ketone group of one with the free aldehyde or ketone group of the other. In the case of sucrose, there is no reducing group present in the resulting compound.

E.g. D-Glucose + Fructose—Sucrose (α1-2 linkage)

For other disaccharides, the union also takes place by fusion of the aldehyde or ketone group of one with an alcoholic group of other, so one aldehyde group remains free and exhibits reduction test, etc.

E.g.

D-Glucose + D-Glucose—Maltose (α1-4 linkage)

D-Galactose + D-Glucose—Lactose (β1-4 linkage)

The three disaccharides discussed are colorless crystalline compounds, sweetish to taste and are important constituents of our daily food. Cane sugar or sucrose is used universally. Maltose is present in sprouted seeds. It is also the end product of salivary and pancreatic digestion of starchy food. Lactose is the milk sugar.

Lactose

Milk sugar - dimer of β-D-galactose and β-D-glucose

O-β-D-Galactopyranosyl-(1→ 4)-β-D-Glucopyranose

Lactose

Lactase – Enzyme required to hydrolyze lactose.

Lactose intolerance –

Lack or insufficient amount of the enzyme. If lactose enters lower GI, it can cause gas and cramps.

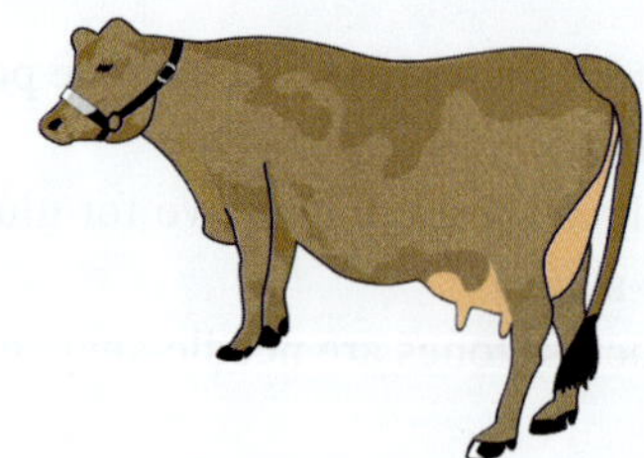

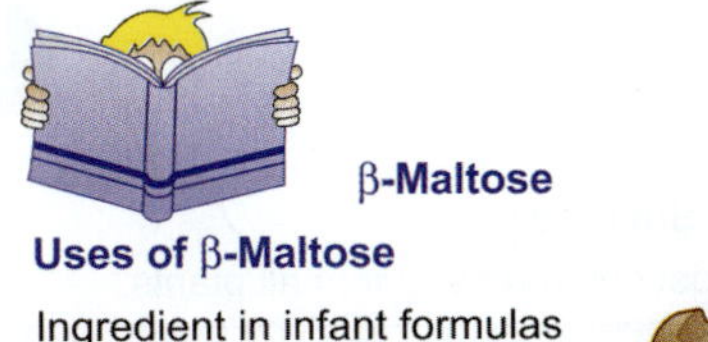

β-Maltose

Uses of β-Maltose

Ingredient in infant formulas
Production of beer
Flavoring-fresh baked aroma

Maltose + $H_2O \rightarrow 2$ glucose.

Lactose and Maltose

Perform the following tests with 2% maltose and 2% lactose solutions.

Expt. 1. Molisch's Test
Expt. 2. Benedict's Test
Expt. 4. Barfoed's Test
Expt. 9. Osazones Test

- Both answer Molisch test positive.
- Both give positive Benedict's test.
- Barfoed's solution is not reduced by any of these sugars because Barfoed's test is in acidic medium and disaccharides being poor reducing agents cannot reduce in acidic medium.

(***Note:*** The test can be used to differentiate mono and disaccharides. It must, however, be remembered that acid present in Barfoed's reagent on prolonged boiling with disaccharides can produce hydrolysis of the disaccharide to its monosaccharide units which can reduce the reagent. Hence the boiling period should not be more than one minute.)

- It is very difficult to get crystals from disaccharides. Unless the instructions already given are adhered strictly, the crystals may take quite a long time to form even under best conditions.

Lactosazones are hedgehog shaped or powder puff shaped. Maltosazones are sunflower shaped.

Sucrose

Sucrose is the sweetest of the three disaccharides. It is a non-reducing sugar as both the functional groups are involved in bond formation.

Sucrose

Table sugar - Most common sugar in all plants.
- Sugar cane and beet are up to 20% by mass sucrose.
- Disaccharide of α-glucose and β-fructose.

O-α-D-glucopyranosyl-(1 ⟶ 2)-β-D-fructofuranoside

Perform the following tests with 1% sucrose:

Expt. 1. Molisch's Test
Expt. 2. Benedict's Test
Expt. 11. Acid Hydrolysis of Sucrose (Specific test for sucrose)

Flow Chart for Identification of Unknown Carbohydrate Solution

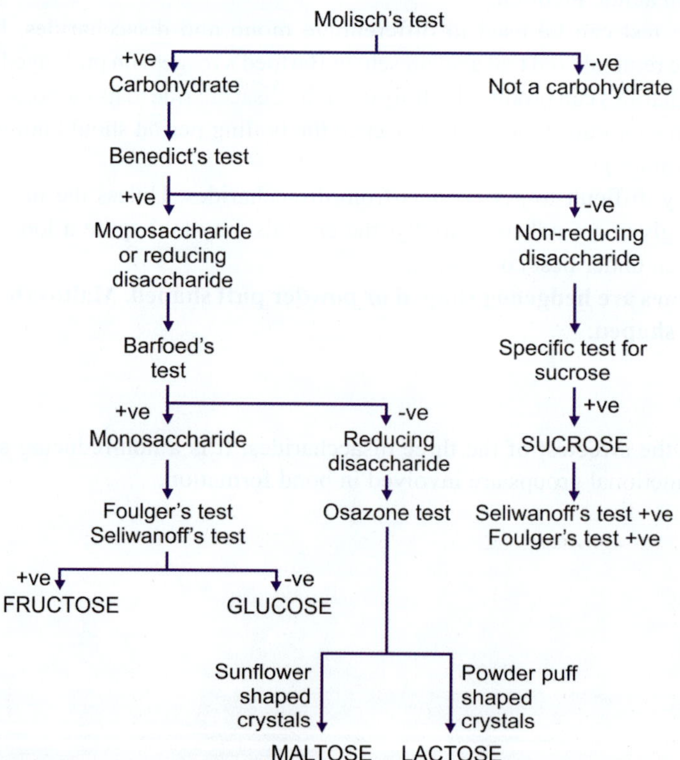

RECAP OF TESTS FOR CARBOHYDRATESS

	Molisch Test	***Benedict's Test***	***Barfoed's Test***	***Foulger's Test***	***Seliwanoff's Test***	***Osazone Test***
REAGENT COMPOSITION	α- naphthol Conc H_2SO_4	$CuSO_4$, Na_2CO_3, Na-citrate	$Cu(AcO)_2$ AcOH	Urea, $SnCl_2$, 40% H_2SO_4	HCl, Resorcinol	Phenyl hydrazine, sodium acetate, Glacial acetic acid.
MEDIUM	Acidic	Alkaline	Weak acidic	Acidic	Acidic	Weak acidic
INTERMEDIATE FORMED	Furfurals	Enediols	Enediols	Furfurals	Furfurals	Osazones
COLOR OBSERVED/ END RESULT	Violet ring at the junction of 2 layers	Brick red ppt	Red scum	Deep blue color	Cherry red color	Yellow crystals
SPECIFICITY OF THE TEST	+ve for all carbohydrates	+ve for all reducing sugars	Distinguishes disaccharides from monosaccharides	+ve for ketose sugars	+ve for ketose sugars	Needle shaped – glucose Sunflower shaped- maltose, Powder puff- lactose.
PRINCIPLE	Dehydration followed by formation of colored complex	Reduction	Reduction	Dehydration followed by formation of colored complex	Dehydration followed by formation of colored complex	Formation of hydrazones followed by osazones

RECAP OF REACTIONS OF CARBOHYDRATES

Sugar	*Classification*	*Molisch Test*	*Benedict's Test*	*Barfoed's Test*	*Seliwanoff's Test*	*Foulger's Test*	*Osazone Test*
GLUCOSE	Monosaccharide	Violet ring +ve	Brick red ppt +ve	Red scum +ve	No cherry red color -ve	No deep blue color -ve	Needle-shaped crystals
FRUCTOSE	Monosaccharide	+ve	+ve	+ve	+ve	+ve	Needle-shaped crystals
MALTOSE	Disaccharide	+ve	+ve	-ve	Not performed	Not performed	Sunflower shaped crystals
LACTOSE	Disaccharide	+ve	+ve	-ve	Not performed	Not performed	Powder puff-shaped crystals
SUCROSE	Disaccharide	+ve	-ve after hydrolysis +ve	Not performed	+ve	+ve	---

Principle: Sucrose on hydrolysis by conc. HCl is converted into glucose and fructose (the reducing monosaccharides) which answer the Benedict's test.

Procedure: To 5 ml of sucrose solution, add 3 drops concentrated HCl and boil for 2 minutes. This will hydrolyze cane sugar to its constituent units glucose and fructose. Now neutralize the solution with 20% sodium carbonate using litmus paper as indicator (this is essential as Benedict's test requires alkaline medium). Boil 1 ml of the neutralized solution with 5 ml Benedict's solution and note the reduction produced.
Expt. 6. Foulger's Test
Expt. 7. Seliwanoff's Test

Sucrose Solution

- Answers positive Molisch's test
- Negative Benedicts test (non-reducing sugar)
- Positive Benedict's test after acid hydrolysis
- Positive Foulger's and Seliwanoff's tests.

RELEVANT QUESTIONS–CARBOHYDRATES

1. Define and classify carbohydrates.
2. What are the functional groups present in carbohydrates?
3. What is the general test for all carbohydrates? What is its principle?
4. What is the composition of Molisch's reagent?
5. What is meant by reducing sugar ? Give example.
6. What are the compositions of Benedict's reagent? What are the functions of each?
7. What is the red precipitate formed in Benedict's test called?
8. What is the difference between glucose and fructose?Which tests will differentiate between the two?
9. Why sucrose is a non-reducing sugar?
10. Name the monosaccharides present in a) Maltose b) Lactose c) Sucrose.
11. What is the other names for sucrose and why is it called so?
12. What are the advantages of Benedict's test over Fehling's test ?
13. What are the components of Barfoed's reagent.
14. Name the test which helps to differentiate reducing monosaccharides from reducing disaccharides.
15. What are the differences between Benedict's and Barfoed's tests?
16. What are the components of a a) Seliwanoff's reagent b) Foulger's reagent?
17. How will you differentiate maltose and lactose?
18. Why does glucose and fructose form the same osazone?
19. What is the confirmatory test for sucrose?
20. Why sucrose cannot form osazones?
21. Why Benedict's test is preferred to Fehling's for detecting reducing sugars in urine?

PROTEINS

INTRODUCTION

Proteins are complex macromolecules **made up of amino acids** produced by all living cells. They have definite size, shape and charge. They are composed of 20 amino acids in varying number and sequences. Amino acids contain both a **carboxylic group** (–COOH) and an **amino group** ($-NH_2$) and a **side chain** (–R Group).The sequence of amino acids determines the characteristics of a protein, which is regulated by genetic code. All proteins contain C, H, O and N (about 16%); some contain S. The presence of N differentiates proteins from carbohydrates and lipids. The proteins perform various functions in the form of **enzymes, hormones, antibodies, coagulation factors, contractile elements** and maintenance of **osmotic pressure.**

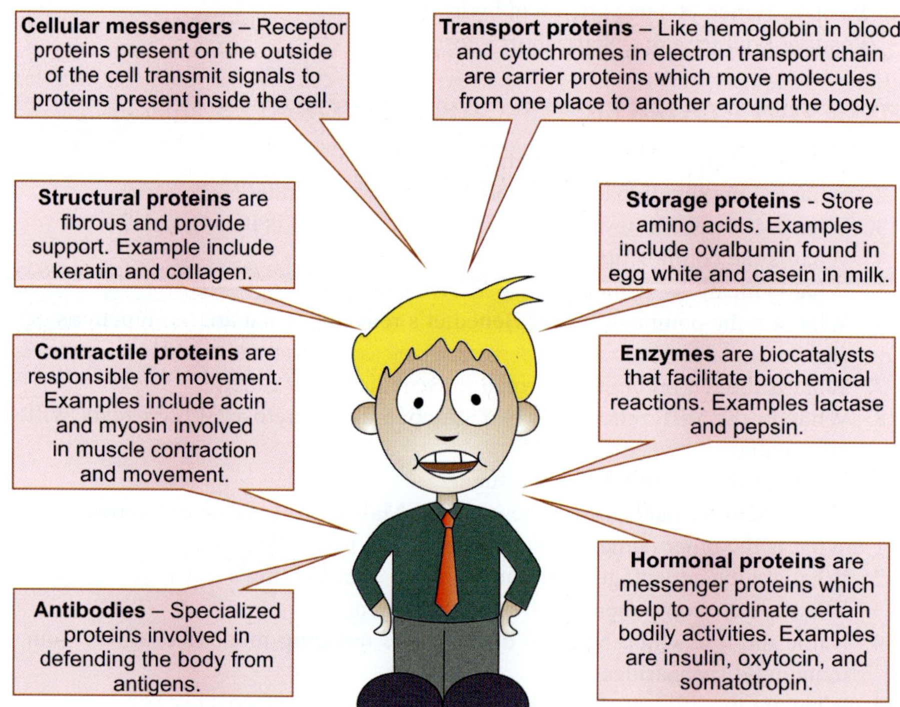

Fig. 2.2: Functions of proteins

GENERAL REACTIONS OF PROTEINS

Reactions of amino acids and proteins can be studied under two broad headings:

A. Precipitation Reactions of Proteins

B. Color Reactions of Proteins

A. PRECIPITATION REACTIONS OF PROTEINS

Solubility of proteins depends on the proportion and distribution **of polar hydrophilic groups** and **non-polar hydrophobic groups** in molecules.

Proteins form colloidal solutions of the emulsion type. A certain amount of water is imbibed by the protein molecule (water of hydration) and is taken up by the various free COOH, NH_2 and other groups.

The stability of the colloidal solution depends on this water of hydration and on the electric charges carried by the molecules. Thus, removal of the water of hydration by any dehydrating agents (E.g. Neutral salts and alcohol) will convert the emulsoid or lyophilic colloid to a suspensoid (Lyophobic colloid). Addition of electrolytes will discharge the protein molecule in suspension and thus precipitate them.

1. PRECIPITATION OF PROTEINS BY HEAVY METAL

Principle: The protein molecules exist as anions (negatively charged) in alkaline medium. The negatively charged protein molecule will combine with the positively charged metallic ions like Ag^{+}, Cd^{+2}, Cu^{+2}, Fe^{+3}, Hg^{+2}, Pb^{+2}, Sb^{+3}, and Zn^{+2} and get precipitated.

An advantage of this type of precipitation is that they can precipitate in dilute solutions as well. These ions are classified into three groups as follows:

i. Bind strongly to carboxylic acids and nitrogenous compounds
   Mn^{2+}, Fe^{2+}, Co^{2+}, Ni^{2+}, Zn^{2+}
ii. Bind to carboxylic acids
   Ca^{2+}, Ba^{2+}, Mg^{2+} and Pb^{2+}
iii. Bind strongly to sulfhydryl groups
   Ag^{+}, Hg^{+} and Pb^{2+}

12. Mercuric Nitrate Precipitation Test

Procedure: To 3 ml of protein solution, add 5 drops of 5% mercuric nitrate solution. Note the white precipitate of mercury proteinate.

Protein precipitation by heavy metals is used for its disinfectant properties in external applications. $AgNO_3$ is used to prevent gonorrhea infections in the eyes of newborn infants. Silver nitrate is also used in the treatment of nose and throat infections, as well as to cauterize wounds.

Mercury salts administered as Mercurochrome or Merthiolate react with the sulphydryl group and precipitate the proteins thereby preventing infections in wounds.

13. Zinc Sulfate Precipitation Test

To 3 ml of protein solution, add 2 drops of 2% zinc sulfate solution. A white precipitation of zinc proteinate is formed.

This property of precipitation of proteins by heavy metals is utilized for the household treatment of accidental ingestion of heavy metals. The protein, egg white or milk acts as an antidote and binds with the heavy metals and prevents absorption of heavy metals.

2. PRECIPITATION BY ALKALOID REAGENTS

Principle: Proteins are amphoteric in nature, i.e. they behave as acids in an alkaline medium and as bases in an acidic medium. In the presence of alkaloid reagents, they act as bases, and react with the acid to form an insoluble salt, e.g. protein sulphosalicylate.

14. Precipitation by Esbach's Reagent

To 2 ml of the protein solution in a test tube, add 2 ml of Esbach's reagent (1 g of picric acid and 1 g of citric acid in 100 ml of water). Note the yellow precipitate of protein picrate.

15. Precipitation by Sulphosalicylic Acid

To 2 ml of the protein solution in a test tube, add a drop of 20% sulphosalicylic acid solution. Note the white precipitate of protein sulphosalicylate.

3. PRECIPITATION OF PROTEINS BY ORGANIC SOLVENTS

Principle: Addition of organic solvents, such as acetone or alcohol decreases dielectric constant of solvent and displaces some water molecules associated with the protein and decreases the concentration of water in the solution. These effects tend to decrease solubility of protein.

16. Precipitation by Ethanol

To 1 ml of protein solution, add 2 ml ethanol. Mix and let it stand for 5 minutes. A white precipitate is formed.

A 70% alcohol solution is used as a disinfectant on the skin. This concentration of alcohol is able to penetrate the bacterial cell wall and denature the proteins and enzymes inside the cell.

4. PRECIPITATION BY NEUTRAL SALTS

Principle: The colloidal protein has electric charge and the shell of hydration to keep it stable in solution. Both these factors are removed by adding neutral salts like ammonium sulfate. The abundance of the salt ions decreases the solvating power of the salt ions, thereby decreasing the solubility of the proteins and results in precipitation.

At low concentrations of salt, solubility of the proteins usually increases slightly (salting in). But at high concentrations of salt, the solubility of the proteins drop sharply (salting out).

Depending on the surface area of the protein molecule, quantity of salt required will vary. Smaller molecules like albumin have relatively large surface area thus require full saturation for precipitation.

Casein and gelatin have smaller surface area and get precipitated by half saturation. Peptones are very small molecules, thus have very large surface area and do not get precipitated even with full saturation.

17. Half Saturation Test

Procedure: To 3 ml of given solution, add equal quantity of saturated ammonium sulfate solution. Mix thoroughly. Keep for 5 minutes. Note the precipitation. Filter the solution.

Test a portion of the filtrate for protein by doing biuret test using 2 volumes of 40% NaOH as above. It is positive. The filtrate contains albumin, which is soluble in half saturated ammonium sulfate solution.

Note: 40% NaOH is added to overcome the interference by ammonium ions which can obscure the violet color of protein by forming deep blue cuprammonium ions [$Cu(NH_3)_4^{++}$]

18. Full Saturation Test

Principle: Refer half saturation test.

Procedure: To the remaining filtrate add solid ammonium sulfate in excess and mix well. Note the formation of precipitate. This consists of albumin, which is insoluble in saturated ammonium sulfate solution.

Filter and test the filtrate for protein by biuret reaction (using 2 volumes of 40% NaOH). It is negative hence all the proteins in solution are precipitated by full saturation with ammonium sulfate.

Note: Substance which disrupts the three dimensional structure in macromolecules such as proteins, DNA, or RNA and denatures them are called chaotropic agent. They disrupt the inter-molecular interactions mediated by non-covalent forces such as hydrogen bonds, van der Waals forces, and hydrophobic effects.

5. PRECIPITATION AT ISOELECTRIC pH

19. Isoelectric Precipitation of Casein

Principle: Fractional precipitation can be achieved by changing the pH of the medium. At acidic pH (lower pH) proteins have a net positive charge because the amide gains an extra proton. At alkaline pH (higher pH) they are negatively charged because the carboxyl group loses its proton. At their iso-electric pH (pI) a protein has no net charge. The solubility of the proteins is the minimum at their iso-electric pH as the protein molecules become electrically neutral at this pH and is unable to interact with the medium and will eventually fall out of the solution. Most proteins can be precipitated by heating them at their iso-electric pH. Casein, however, is peculiar in that it is precipitated at its iso-electric pH (4.6) even at room temperature.

Procedure: To 5 ml of the solution add a drop of bromocresol green indicator (yellow at pH 3.8 and green at pH 5.4). Continue addition of 1% acetic acid and mix gently till maximum precipitation is obtained at the iso-electric pH of casein.

6. PRECIPITATION BY MINERAL ACIDS

20. Heller's Test

Principle: Proteins are denatured by nitric acid with the formation of a white precipitate at the junction of the two layers (This differs from the nitration reaction in "xanthoproteic acid test").

Procedure: Take 2 ml of conc. Nitric acid in a test tube. Incline the tube and add, by using a pipette, equal volume of the protein solution so that the latter forms a layer on the surface of the acid.

A white ring appears at or immediately below the junction of the two fluids. The precipitate does not dissolve on gentle warming.

Denaturation is a process whereby the secondary, tertiary and quaternary structures of proteins are lost, retaining its primary structure. Denaturation may be brought about both by physical factors, such as high temperature (above 60°C), or chemical compounds, like: acids, bases, ions of heavy metals, urea, detergents, or some organic solvents. Denaturation changes some physical and chemical properties of proteins, e.g. it alters the solubility of some proteins or increases the susceptibility of some proteins to the action of proteolytic enzymes. In most cases the denaturation is an irreversible process. If the action of denaturing agent was mild and for a short time the denaturation may be reversed. Such a process is named renaturation. Renaturated protein regains its natural structure and properties.

Some proteins form lyophilic colloid solutions. The stability of colloids depends mainly on electric charge of protein molecules, their hydration and temperature of protein solution. Proteins tend to lose the lyophilic character at isoelectric point under the action of denaturing agent and precipitates from the solution. Such a process is named coagulation. Coagulation is an irreversible process.

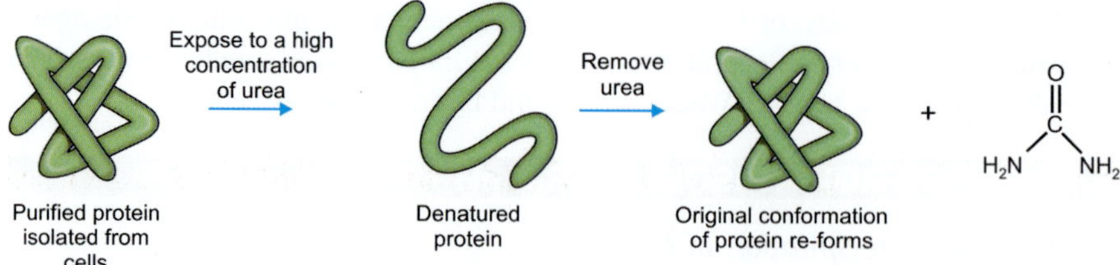

B. COLOR REACTIONS OF PROTEINS

The functional groups in amino acids and proteins can react with compounds to produce characteristically colored products. The intensity of the color formed by a particular group varies among proteins in proportion to the number of reacting functional or free groups present and their accessibility to the reacting reagents.

21. Biuret Reaction

Principle: Cupric ions in an alkaline medium form a violet colored complex with peptide bond nitrogen of peptides and Proteins.

This is a test for peptide linkages. Since all proteins contain peptide linkages, they respond to this test. The violet color is due to the formation of copper co-ordination complex between cupric hydroxide and peptide bond.

The tests may also be given by substances containing (– $CSNH_2$), (– $C(NH)NH_2$), and (– CH_2NH_2) groups.

Biuret, which is obtained by heating urea at 180°C, also gives a positive test as it contains CONH – linkages. In fact, the name of the test is based on this. Minimum requirement for a positive test is the presence of two peptide bonds.

$$2CO(NH_2)_2 \rightarrow H_2N\text{-}CO\text{-}NH\text{-}CO\text{-}NH_2 + NH_3$$

O O

H_2N N NH_2

H

Biuret

Procedure: To 2 ml of protein solution add an equal volume of 5% NaOH and a drop or two of 1% copper sulfate solution and mix. A violet or purple color is observed.

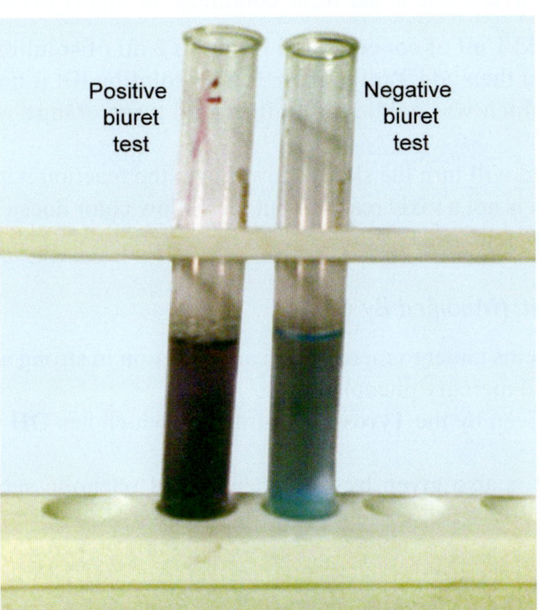

Note: (i) Peptone gives a pink color, and may require the addition of a larger amount of copper sulfate solution than usual. (ii) Gelatin may give a very faint color and, therefore, a control may be run to avoid confusion.

22. Ninhydrin Test

Principle: Ninhydrin reacts with α-amino groups of proteins and free amino acids to give a blue or purple colored complex. This is one of the most sensitive **tests for amino acids** answered by all **proteins, peptones, peptides, amino acid and ammonia.**

α-Amino acid + Ninhydrin $\longrightarrow$ Aldehyde + Hydrindantin $CO_2 + NH_3$
Hydrindantin + NH_3 + Ninhydrin $\longrightarrow$ Blue complex

Procedure: To 1 ml of protein solution, add 2 drops of 0.2% ninhydrin solution and boil for 2 minutes. A blue color indicates the presence of proteins in the solution. This test is given by all solutions which contain at least one free amino group and one carboxyl group. Hence, this test is also given by amino acids in addition to proteins and their derivatives.
Note: Amino acid Proline which has an imino group will produce an orange color instead of the purple-blue color.

23. Xanthoproteic Reaction

Principle: On heating with conc. HNO_3, proteins containing aromatic amino acids form yellow color due to the nitration of the **benzene ring.** The nitro compound is freely ionized in alkaline medium and thus intensifies the color to orange by adding strong alkali; **Tyrosine and Tryptophan** are responsible for this color reaction. Nitration of phenylalanine under these conditions normally does not take place.

Procedure: Add 1 ml of concentrated HNO_3 to 2 ml of solution and boil. Cool the solution and then add 2 ml of 40% NaOH solution till it becomes alkaline. The solution which was acidic with nitric acid turns orange when it becomes alkaline.
Note: Nitric acid will turn the skin yellow due to the reaction with phenyl groups in the skin. This is not a toxic reaction but the yellow color doesn't disappear until the skin cells are replaced.

24. Millon's Test: (Modified By Cole)

Principle: Proteins undergo mercuration and nitration in strong acidic medium to form red colored mercury phenolate.

The test is given by the **Tyrosine amino** acid which has **OH in the benzene ring**.
Note: This test is also given by free phenols and phenolic substances such as salicylic acid.

Excess of chloride interferes by leading to the formation of unionized mercuric chloride. If so, add more of mercuric sulfate.

Procedure: To 2 ml of protein solution, add 2 ml of acid mercuric sulfate solution (10% mercuric sulfate in 10% sulfuric acid). The protein gives yellow precipitate of mercury protein complex which adheres to the sides of the test tube. Boil the solution for 1 minute and cool under tap. Then add 5 drops of 1% sodium nitrite and warm gently. The precipitate or solution turns into red color.

25. Aldehyde Test (Hopkins-Cole-Adamkiewicz Reaction)

Principle: The **indole ring of tryptophan** combines with aldehydes, e.g formaldehyde in the presence of conc. Sulfuric acid to form a violet colored compound. Gelatin gives a negative test as it does not contain tryptophan.

Procedure: To 1 ml of protein solution, add 1 drop of 1 in 500 commercial formalin, (40% formaldehyde). Add 1 drop of 10% mercuric sulfate in 10% H_2SO_4 and mix thoroughly. Then add 1 ml concentrated sulfuric acid along the side of the test tube. It forms a separate layer and a deep violet or purple ring forms at the junction of the two layers.

26. Pauly's Test (Test for Histidine in the Absence of Tryptophan)

Principle: Diazotized sulphanilic acid reacts with the **imidazole group of Histidine** in alkaline medium to form cherry red complex.

Procedure: To 5 ml of solution, add 1ml of 1% sulphanilic acid in 10% HCl, followed by 1 ml of 5% sodium nitrite solution. Mix and allow to stand for 3 minutes. Make alkaline with 3 ml of 10% (Na_2CO_3) solution, add 5 ml ethanol. Mix and cool under tap water. A permanent red color indicates the presence of histidine.
Note: Tryptophan gives similar color.

27. Sakaguchi's Test

Principle: In an alkaline medium, α-naphthol combines with the **guanidine group of arginine** to form a complex which is oxidized by sodium hypobromite to produce a red color.

Procedure: To 2 ml of solution, add 2 drops of 40% NaOH and 2 drops of 1% alpha naphthol in alcohol. Add 10 drops of bromine water. Mix the solution well. Note the carmine red color of the solution. The test is specific for arginine.

The reagent always gives certain amoumt of color. A control is indispensable. Ammonia interferes with the test by reacting with sodium hypobromite in large amounts.

28. Lead Sulfide Test (Sulfur Test)

Principle: On boiling with NaOH, the sulfur present in the **sulfur containing amino acids (Cysteine, Cystine)** of protein is liberated as sodium sulfide. This reacts with lead acetate to form a brown or black precipitate of lead sulfide.

Procedure: Take 2 ml of solution and add 2 ml of 40% NaOH. Boil the solution for 1 minute and cool. Add 2 drops of 2% lead acetate solution. A brown or black color appears due to the formation of lead sulfide.

This test is not given by methionine because it has a strong thioether linkage and sulfur is not converted to sulfide.

REACTIONS OF ALBUMINS AND GLOBULINS

Introduction

Albumins and globulins are the proteins found in plasma, milk and egg white. They are simple proteins, soluble in water and dilute salt solutions.

They are coagulated by heating and are precipitated on full saturation and half saturation respectively. They are complete proteins having all the essential amino acids.

Expt. No. 21. Biuret test

29. Heat Coagulation Test

Principle: When a protein is heated, its physical, chemical and biological properties are changed due to breaking up of certain bonds and the resultant change in the conformation of its molecules. This process is known as denaturation.

However, when the **coagulable proteins** such as **albumin and globulin** are heated at their isoelectric pH, a series of changes occur involving dissociation of the protein subunits (disruption of quaternary structure), uncoiling of the polypeptide chains (disruption of tertiary and secondary structure) and matting together of the uncoiled polypeptide chains (coagulation).

While a denatured protein may be restored to its original structure and function by certain manipulations, coagulation is an irreversible process.

Procedure: Fill two-thirds of the test tube with protein solution and add one or 2 drops of chlorophenol red indicator, (shows yellow color at pH 4.8 and below; red at 6.4 and above). A solution giving a light pink color with this indicator indicates a pH of 5.4. Adjust the pH to 5.4 by adding a drop or two of 1% acetic acid or 1% sodium carbonate as the case may be. Heat the top layers of the solution and note the appearance of dense coagulum, the lower half acts as a control.

Medical supplies and instruments are sterilized by heating to denature proteins in bacteria and thus destroy the bacteria

TABLE 2.1: Methods of protein denaturation

Denaturing action	Mechanism of operation
Heat	Hydrogen bonds are broken by increased translational and vibrational energy (Coagulation of egg white albumin on frying)
Ultraviolet radiation	Similar to heat (sunburn)
Strong acids or bases	Salt formation; disruption of hydrogen bonds (skin blisters and burns, protein precipitation)
Urea solution	Competition for hydrogen bonds (precipitation of soluble proteins)
Some organic solvents (e.g. ethanol and acetone)	Change in dielectric constant and hydration of ionic groups (disinfectant action and precipitation of protein)
Agitation	Shearing of hydrogen bonds (beating egg white albumin into a meringue)

17. Half Saturation Test

Principle: Proteins, which are colloidal in nature, are kept in solution by two factors:

1. **Electric charges**: A large number of electric charges are present on the surface of protein molecules. The similarly charged particles repel each other and this prevents their coalescence.
2. **Shell of hydration**: Each molecule is surrounded by a film of water known as the shell of hydration. The shell of hydration also prevents coalescence of particles.

If both these factors are removed, the particles coalesce, and are precipitated. This can be done by adding a neutral salt such as ammonium sulfate which neutralizes the electric charges, and removes the shell of hydration as it has got greater affinity for water than the colloid.

The amount of ammonium sulfate required to precipitate a colloid depends upon the surface area of the particles. The larger the surface area, the greater the number of electric charges and the larger the shell of hydration.

Thus small molecules, e.g. **albumin,** having a relatively larger surface area are **precipitated** only by **full saturation with ammonium sulfate**.

The larger molecules, e.g. **globulin** , casein and gelatin, have a smaller surface area and are, therefore, **precipitated** both by **half saturation** and **full saturation with ammonium sulfate.**

Peptone, which has got very small molecules is not precipitated even by full saturation with ammonium sulfate.

The process of precipitation is called as **"salting out"**.

Procedure: To 3 ml of given solution, add equal quantity of saturated ammonium sulfate solution. Mix thoroughly. Keep for 5 minutes. Note the precipitation. Filter the solution.

Test a portion of the filtrate for protein by doing biuret test using 2 volumes of 40% NaOH as above. It is positive. The filtrate contains albumin, which is soluble in half saturated ammonium sulfate solution.

Note: 40% NaOH is added to overcome the interference by ammonium ions which can obscure the violet color of protein by forming deep blue cuprammonium ions [Cu $(NH_3)_4^{++}$].

18. Full Saturation Test

Principle: Refer Half Saturation Test

Procedure: To the remaining filtrate, add solid ammonium sulfate in excess and mix well. Note the formation of precipitate. This consists of albumin, which is insoluble in saturated ammonium sulfate solution.

Filter and test the filtrate for protein by he biuret reaction (using 2 volumes of 40% NaOH). It is negative hence all the proteins in solution are precipitated by full saturation with ammonium sulfate.

Expt. no. 24. Millons test
Expt. no. 25. Aldehyde test
Expt. no. 28. Sulfur test and Lead sulfide test
Expt. no. 1. Molisch's test

Albumins and Globulins observation:

- Biuret test – Positive as both are proteins.
- Heat coagulation test – Positive for both Albumin and Globulin as both are coagulable proteins.
- Half saturation test –Positive for Globulin (biuret is also positive as albumins are in the filtrate), since they have smaller surface area they get precipitated.
- Full saturation test –Positive for Albumin (Biuret is negative) as they have larger surface area and get precipitated.
- Millon's test, Aldehyde test and Sulfur test – Positive as Albumin and Globulin are complete proteins.
- Molisch test – Positive as Albumin and Globulin are glycoproteins

Demulcents like egg albumin, starch or milk act as physical or mechanical antidote and prevent the action of poison. Egg albumin also precipitates alkaloids and metal poisons like mercury, arsenic and other heavy metals.

REACTIONS OF GELATIN

Introduction

Gelatin is a derived protein formed from the collagen, connective tissue protein, by boiling with water. Conversion involves hydrolysis of some of the covalent bonds of collagen. This is the major reason for cooking meat, since it is the collagen of connective tissue that makes meat tough.

While collagen is insoluble in water and resistant to animal digestive enzymes, gelatin is easily digestible and soluble. If the solution is concentrated it sets into a gel on cooling hence the name gelatin. This property of gelatin is used in preparation of ice creams and jelly. It is also used in pharmaceutical industry for preparation of capsules.

Collagen is very rich in glycine, alanine, proline and hydroxyproline and is low in nearly all the other amino acids. It is deficient in tryptophan, cysteine and Cystine. Gelatin is precipitated by half saturation as it has a high molecular weight.

Gelatin can be heated in water to carry out all the tests. Nutritionally, it is a poor quality protein.

Expt no. 21. Biuret test

Expt no. 29. Heat coagulation test

Expt no. 17. Half saturation test followed by biuret test

To 2 ml of Gelatin solution, add 2 ml of saturated Ammonium sulfate solution (This becomes ½ saturation with Ammonium sulfate). Mix thoroughly, wait for 5 minutes. Filter it. Perform the Biuret test with the filtrate. This Biuret test should be performed by using 40% NaOH, double the volume of the filtrate.

The test is negative indicating that gelatin is completely precipitated by ½ saturation with Ammonium Sulfate.

Expt no. 24. Millon's test

Expt no. 25. Aldehyde test

Expt no. 28. Lead sulfide test

Gelatin observation:

- Biuret test – Positive as Gelatin is a protein.
- Heat coagulation test – Negative, gelatin is a non-heat coagulable protein.
- Half saturation test – Positive (Biuret is negative), Gelatin has smaller surface area.
- Millon's test – Faintly positive, contains very little tyrosine.
- Aldehyde test and sulfur test – Negative, tryptophan and sulfur containing amino acids – Cysteine and cystine are deficient.

REACTIONS OF PEPTONES

Introduction

Peptones, classified as derived proteins are the partial degradation (hydrolytic) products of proteins. They are relatively low molecular weight compounds.

They are not heat coagulable and are not precipitated even by full saturation with ammonium sulfate. They are often used in the preparation of culture media for growing microorganisms.

Expt no. 21. Biuret test: Peptone gives a pink color, and may require the addition of a larger amount of copper sulfate solution than usual.
Expt. no. 29. Heat coagulation test
Expt. no. 17. Half saturation test followed by Biuret test.
Expt. no. 18. Full saturation test followed by Biuret test.
Expt. no. 28. Lead sulfide test.

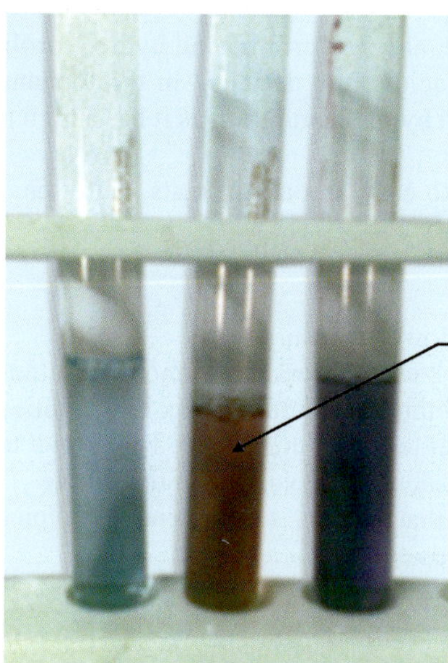

Positive biuret test (pink color for peptone) Purple color for other proteins

30. Precipitation by Phosphotungstic Acid

Principle: Proteins are amphoteric in nature, i.e. they behave as acids in an alkaline medium and as bases in an acidic medium. In the presence of alkaloidal reagents like, sulphosalicylic acid and phosphotungstic acid, they act as bases, and react with the acid to form an insoluble salt, e.g. protein sulphosalicylate, phosphotungstic acid, etc.

Procedure: To 2 ml of peptone solution, add 0.5 ml of phosphotungstic acid. A white precipitate is observed.

Filter the above solution. Perform Biuret test with the filtrate. The biuret will be negative due to precipitation of peptones.
Expt. No. 24. Millon's test
Expt. No. 25. Aldehyde test
Expt. No. 28. Lead sulfide test

Peptones observation:

- Biuret test–Positive, as peptone is a protein.
- Heat coagulation test–Negative, as peptone is a non-heat coagulable protein.

- Half saturation and full saturation tests–negative as Peptones are very small particles having very large surface area.
- Phosphotungtic acid test–Peptones are precipitated by general protein precipitants.
- Millon's and Aldehyde test are positive–Peptones contain tyrosine and Tryptophan.
- Sulfur test–Negative or faintly positive, peptones are deficient in cysteine and cystine.

REACTIONS OF CASEIN

Introduction

Casein is the major **milk protein** present as calcium caseinate. It is a conjugated protein (phosphoprotein) with phosphate group attached to hydroxyl group of serine and threonine residues.The pH of casein is 6.6 and its isoelectric point is 4.6.

This is a non-coagulable protein and is soluble in dilute alkaline and acidic solution. This is precipitated by half saturation with NH_4SO_4.

This protein lacks in sulfur containing amino acids. Casein can be tested for the presence of R groups (except, sulfur containing groups), phosphorus and calcium. Casein is deficient in glycine, serine and cysteine.

Casein is hydrolyzed by pepsin and chymotrypsin into paracasein.

Expt. No. 21. Biuret test

Expt. No. 29. Heat coagulation test

Expt. No. 17. Half saturation test

Expt. No. 19. Isoelectric precipitation of casein.

Principle: The solubility of the proteins is minimum at their isoelectric pH as the protein molecules become electrically neutral at this pH. Most proteins can be precipitated by heating them at their isoelectric pH. Casein, however, is peculiar in that it is precipitated at its isoelectric pH (4.6) even at room temperature.

Procedure: To 5 ml of the solution, add a drop of bromocresol green indicator (yellow at pH 3.8 and green at pH 5.4). Continue addition of 1% acetic acid till maximum precipitation is obtained.
Note: Maximum precipitate occurs at green color and the pH at which the green color appears is the iso-electric pH for casein, pH 4.7.

31. Test for Inorganic Phosphorus

Principle: The organic phosphate present in phosphoprotein casein is converted to inorganic phosphate on boiling with strong NaOH solution. Inorganic phosphate reacts with ammonium molybdate to form ammonium phosphomolybdate which is canary yellow in color.

Procedure: To 3 ml of casein solution, add 0.5 ml of 40% NaOH solution. Heat strongly and cool under tap water. Add 0.5 ml of conc. HNO_3 and filter. To the filtrate, add a pinch of solid ammonium molybdate and warm gently. Note the canary yellow color of precipitate.
Expt. No. 24. Millon's test
Expt. No. 25. Aldehyde test
Expt. No. 28. Lead sulfide test

Casein observation:

- Biuret test–Positive, as Peptone is a protein
- Heat coagulation test–Negative, Casein is a non-heat coagulable protein.
- Half saturation test–Positive (biuret negative), casein are larger molecules having smaller surface area.
- Isoelectric pH precipitation–Positive, casein is precipitated at pI, i.e. pH 4.6.
- Test for inorganic phosphorus–Positive, casein is a phosphoprotein.
- Millon's and Aldehyde tests–Positive, both tyrosine and tryptophan are present in casein.
- Sulfur test –Negative, cysteine and cystine are absent in casein.

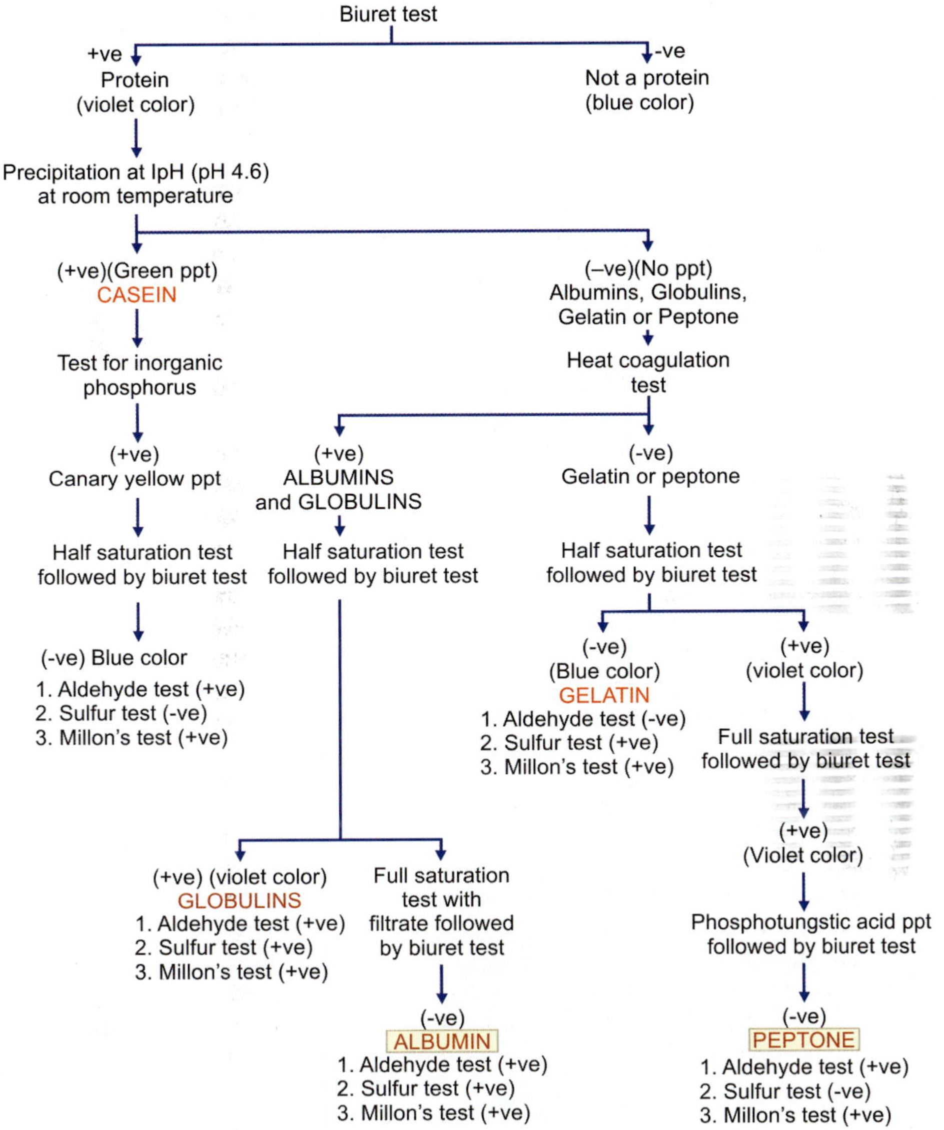
Biuret test
+ve
Protein
(violet color)
-ve
Not a protein
(blue color)
Precipitation at IpH (pH 4.6)
at room temperature
(+ve)(Green ppt)
CASEIN
(–ve)(No ppt)
Albumins, Globulins,
Gelatin or Peptone
Test for inorganic
phosphorus
Heat coagulation
test
(+ve)
Canary yellow ppt
(+ve)
ALBUMINS
and GLOBULINS
(-ve)
Gelatin or peptone
Half saturation test
followed by biuret test
Half saturation test
followed by biuret test
Half saturation test
followed by biuret test
(-ve) Blue color
1. Aldehyde test (+ve)
2. Sulfur test (-ve)
3. Millon's test (+ve)
(-ve)
(Blue color)
GELATIN
1. Aldehyde test (-ve)
2. Sulfur test (+ve)
3. Millon's test (+ve)
(+ve)
(violet color)
Full saturation test
followed by biuret test
(+ve)
(Violet color)
(+ve) (violet color)
GLOBULINS
1. Aldehyde test (+ve)
2. Sulfur test (+ve)
3. Millon's test (+ve)
Full saturation
test with
filtrate followed
by biuret test
Phosphotungstic acid ppt
followed by biuret test
(-ve)
ALBUMIN
1. Aldehyde test (+ve)
2. Sulfur test (+ve)
3. Millon's test (+ve)
(-ve)
PEPTONE
1. Aldehyde test (+ve)
2. Sulfur test (-ve)
3. Millon's test (+ve)

Recap of Reactions of Proteins

	Protein Classification	*Biuret Test*	*Heat Coagulation Test*	*Precipitation at pI*	*Half Saturation Test*	*Full Saturation Test*	*Test for Phosphorus*	*Phosphotungstic Acid Test*	*Millon's Test*	*Aldehyde Test*	*Sulfur Test*
ALBUMINS & GLOBULINS	Simple	+ ve	+ve	-ve	+ve (for globulins)	+ve (for Albumins)	Not done	+ve (Not done)	+ve	+ve	+ve
GELATIN	Derived	+ve	-ve	-ve	+ve	Not done	Not done	+ve (Not done)	Faintly +ve	-ve	Faintly +ve
PEPTONE	Derived	+ve	-ve	-ve	-ve	-ve	Not done	+ve	+ve	+ve	-ve
CASEIN	Conjugated	+ve	-ve	+ve	Not done	+ve	+ve	(Not done)	+ve	+ve	+ve

RELEVANT QUESTIONS–PROTEINS

1. What are the various protein precipitation methods?
2. Name the heavy metals which are used to precipitate proteins.
3. Name the alkaloidal reagents which are used to precipitate proteins.
4. What is the clinical application of precipitation of proteins by alkaloidal reagents?
5. What is the clinical application of precipitation of proteins by heavy metal?
6. What is denaturation? Name few denaturing agents.
7. What is the difference between denaturation and coagulation?
8. What is the clinical application of heat coagulation test?
9. What is isoelectric pH? What is the isoelectric pH of casein?
10. Which is the general color reaction for all proteins?
11. What is the principle of Biuret test?
12. What is the practical clinical utility of Biuret test?
13. What is the general test for all amino acids? What is its application?
14. Is gelatin of high nutritive value? Why?
15. Which amino acid gives positive Hopkins – Cole test?
16. Name the amino acid with indole ring. Name the test to identify the amino acid.
17. Why methionine does not answer sulfur test?
18. Which color reactions are not answered by (a) casein, (b) gelatin and why?
19. Which protein is not precipitated even by full saturation and why?
20. Why does egg albumin answer Molish's test?
21. Will phosphotungstic acid precipitate albumins, globulins and casein from solutions? Why?
22. Why only the top layers of the protein solution is heated in heat coagulation test?

URINE ANALYSIS: WHEN AND WHY

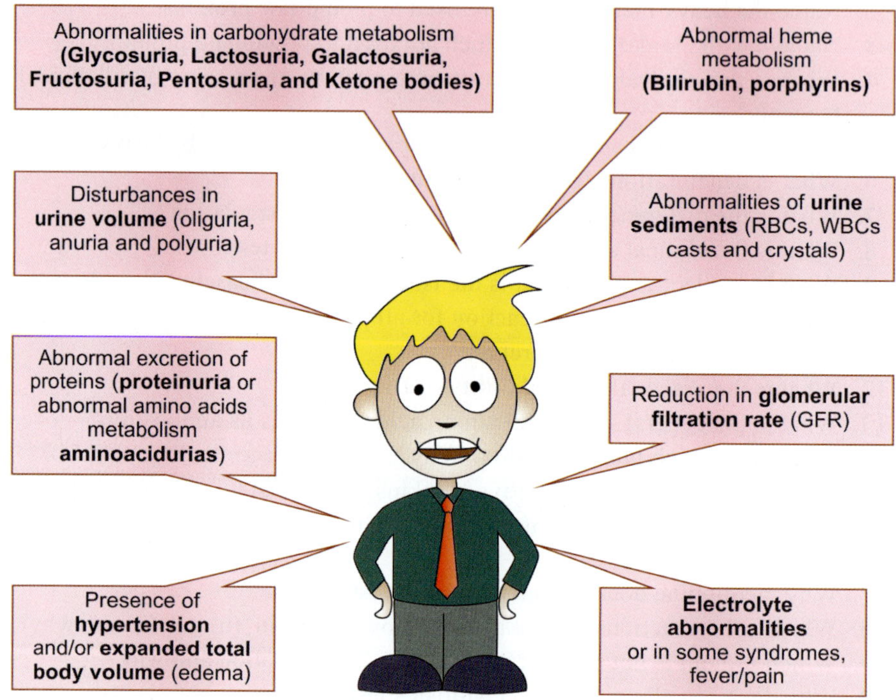

URINE ANALYSIS: NORMAL CONSTITUENTS OF URINE

The clinical laboratory urine examination can provide a wide variety of useful information regarding renal and systemic diseases that may affect this organ. Both structural and functional disorders of the kidney and lower urinary tract may be elucidated as well as the sequential information about the disease, its cause and prognosis.

Careful laboratory examination of urine often narrows the clinical differential diagnosis of numerous renal diseases. Properly performed and interpreted laboratory urine tests remain an essential part of clinical medicine.

NORMAL CONSTITUENTS OF URINE IN A 24-HOUR SAMPLE

Constituents	*In Gms/24 hrs*
Water	1,300 – 1600 ml
Total Nitrogen (Kjeldahl)	10 – 17
Urea	20 – 35
Uric acid	0.1 – 2.0
Creatinine	1.0 – 1.5
Amino acid (free & combined)	~ 1.5
Glucose	~ o.1
Nonglucose reducing substances	~ 1.0
Citric acid	~ 0.3
Ascorbic acid	~ 0.025
Oxalate	~ 0.015
Acetone bodies	~ 0.01
Total sulfur	~ 1.0
Phosphate (as P)	0. 4 – 2.2
Sodium	3 – 6
Potassium	2 – 4
Calcium	~ 0. 2

URINE COLLECTION AND PRESERVATION

Proper urine collection and preservation is important for timely and valid laboratory report.

Types of collection:

1. Random specimen.
2. Mid stream first morning specimen.
3. 24-hour collection specimen

   1. **Random specimen:** The most commonly obtained sample (though not the sample of choice) for biochemical and microscopic analysis. These samples can be collected any time on the spot, therefore, readily available and easy to obtain.

      It is commonly used for glucose, ketone bodies, bile and blood pigments, etc.

   2. **Mid stream first morning specimen.** The specimen of choice for urinalysis as it contains concentrated amount of analytes during its overnight collection in the bladder. The patient is asked to discard the initial part of the first voided urine in the morning and then collect 15 – 20 ml of urine in a clean glass or plastic container.

Urine sample should be analyzed immediately or stored in a refrigerator if analysis is delayed by one or two days

3. **24-hour urine collection:** 24-hour urine collection is required for estimation of urinary creatinine, urea nitrogen, calcium, proteins, glucose, sodium and potassium and certain hormones like catecholamines, 17- hydroxysteroids etc. excretion per day as these have diurnal variations.

   Start collection at 6 AM. Let the patient empty the bladder and discard the first sample. From the second sample onwards, collect into the container (having preservative) all the samples of urine passed till the next morning 6 AM sample, which should be the last sample collected.

Preservatives: The commonly used preservatives for 24 hours collection of urine are – 50 ml of 2 N HCl or 10 ml of conc. HCl per 24 hours collection; or thymol crystals – 5 ml of 100 gms/L solution in isopropanol.

Routine examination of urine is usually conducted under three sections:

a. Physical examination
b. Chemical examination, and
c. Microscopic examination.

Stability of Analytes in Urine

The chemical changes which may occur in urine specimens stored at room temperature include:

- Breakdown of urea to ammonia by bacteria, leading to an increase in the pH of the urine. This may cause the precipitation of calcium and phosphates.
- Oxidation of urobilinogen to urobilin.
- Destruction of glucose by bacteria.
- Precipitation of urate crystals in acidic urine.

These chemical changes can be slowed down by refrigerating the urine at 2–8°C or adding preservatives.

Physical Examination

The physical examination of urine is the initial part of routine urinalysis. This includes assessment of volume, odor, appearance (color and turbidity) and specific gravity.

1. Volume

Volume of urine is influenced by fluid intake, solutes excreted (sodium, urea, etc.), loss of fluid by perspiration and respiration and also cardiovascular and renal status.

Normally an adult excretes between 1000 to 2500 ml urine per day depending upon the amount of fluid intake and amount lost through skin, lungs and bowel.

Volumetric ha for measuring urine volume

Polyuria –Defined as urine volume greater than 2.5 L/day.

Causes:

- Diabetes mellitus,
- Diabetes insipidus
- Later stages of chronic renal failure
- Drugs like diuretics

Oliguria – Defined as urine volume less than 300 ml/day

Causes:

- Fever
- Acute nephritis
- Early stages of chronic glomerulonephritis
- Diarrhea
- Cardiac failure

Anuria – Defined as no urine output in 24 hours.

Causes:

- Shock
- Acute tubular necrosis
- Mercury poisoning
- Incompatible blood transfusion.

2. *Appearance*

- Usually freshly voided urine is clear.
- On standing, phosphate/urate/oxalate crystals may form and make the appearance turbid.
- Abnormally, urine may appear turbid due to the presence of WBC, RBC or bacteria.

3. *Odor*

Normal urine has slightly ammoniacal odor. Other variations commonly seen are:
Fruity odor – Ketoacidosis due to acetone.
Foul smell – Bacterial infection.
Mousy smell – Phenylketonuria.

4. Color

Color of urine is determined to a large extent by its degree of concentration.

Normal urine is pale yellow in color. Other conditions which produce abnormal colors are:

Color	*Metabolite*	*Clinical condition*
Red	RBC Hemoglobin	Hematuria Hemoglobinuria
Red- brown	Myoglobin Porphyrin	Myoglobinuria Porphyrinuria Menstrual-contamination
Yellow	Urochrome	Healthy
Yellow-orange	Urobilin	Dehydration Jaundice
Yellow-green	Bilirubin Biliverdin	Jaundice
Brown-black	Homogentisic acid Methemoglobin porphyrin	Alkaptonuria
Milky-white	Chyle	Chyluria

CONDITIONS IMPARTING ABNORMAL COLOR TO URINE

5. Specific Gravity

Specific gravity is directly proportional to the concentration of solutes excreted. Specific gravity measurement serves as a partial assessment of the kidneys to concentrate urine.

Specific gravity of normal urine may lie between 1.012 to 1.024 depending upon whether urine is dilute (when large volumes are passed) or concentrated.

Fixed specific gravity of 1.010 is suggestive of CRF.

Specific gravity is detected by Urinometer.
High specific gravity:
Conditions:
- Restricted water intake, dehydration
- Presence of glucose in urine (Diabetes mellitus)
- Presence of protein in urine (Proteinuria)
- Adrenal insufficiency.

Low specific gravity:
Conditions:
- Polyuria (except diabetes mellitus)
- High fluid intake
- Diabetes insipidus
- Hypothermia

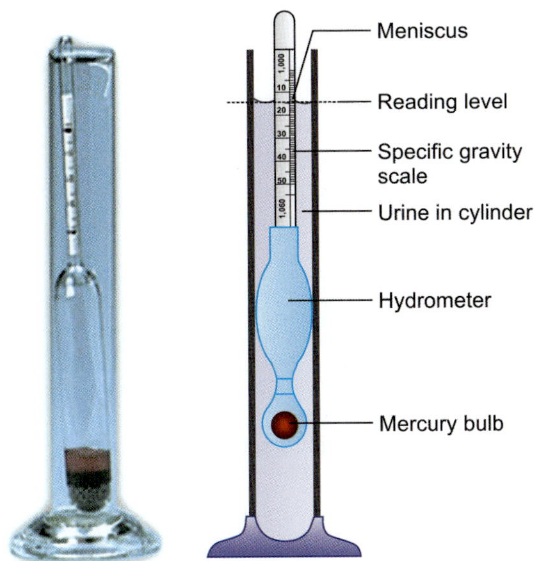

Fig. 2.3: Urinometer

Fixed specific gravity of 1.010: In severe renal diseases **(chronic renal failure)** urine is produced with a fixed specific gravity that is identical to that of the glomerular filtrate, approximately 1.010.

In cases of pronounced glucosuria or proteinuria **correction factor** can be used to adjust the specific gravity to a more accurate value:

- For every 1 gm/dl of **glucose**, 0.004 is subtracted.
- Similarly for every 1 gm/dl of **protein**, 0.003 is deducted.
- **Temperature correction:** Urinometer is graduated at 15°C (60°F) or 20°C (66°F). For each 3°C difference, 0.001 must be added if above or subtracted if below the calibration temperature.

6. Reaction

Freshly voided urine is usually acidic to litmus but may be neutral or faintly alkaline. On standing it may become alkaline due to the formation of ammonia from bacterial decomposition. In healthy individuals, the urine excreted is alkaline particularly after meals as there is secretion of HCL in the gastric juice and thereby compensates for the rise in blood pH. This phenomenon is called 'alkaline tide' urine pH may vary from 4.6 to 8.

The pH of urine is also influenced by the diet.

- High protein or low carbohydrate diet produces acidic urine
- Diet rich in vegetables and fruits produces alkaline urine

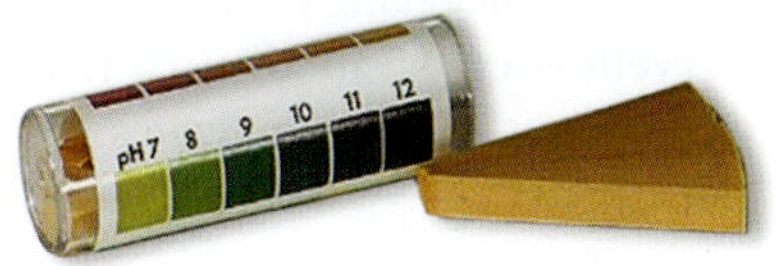

pH paper

Acidic urine:

- High protein diet
- Starvation
- Dehydration and diarrhea
- Diabetic ketoacidosis
- Metabolic and respiratory acidosis

Alkaline urine:

- Vegetable diet
- Vomiting
- Renal tubular acidosis
- Respiratory and metabolic alkalosis
- Ammonia producing, urea splitting bacteria
- Chronic renal failure
- Low carbohydrate diet

CHEMICAL EXAMINATION

Urine may be investigated either for its normal constituents or for abnormal constituents. The normal constituents can be broadly divided into organic and inorganic constituents.

Several enzymes are present in urine; notable amongst them are amylase and trypsin.

It is to be remembered that some of the normal constituents of urine are excreted in abnormal (increased or decreased) amount in certain pathological conditions.

ORGANIC CONSTITUENTS OF NORMAL URINE

The chief organic non-protein nitrogenous constituents (NPN) are urea, uric acid, creatinine, ammonia and amino acids.

Urea

Urea is a diamide of carbonic acid represented by the formula $CO(NH_2)_2$. Urea is soluble in water and alcohol but not in ether or chloroform.

Urea is the chief end product of protein metabolism and is formed in the liver. Urea is nontoxic even when present in relatively large amounts in blood. High blood urea levels indicate inadequate excretory function. Urea is a diuretic.

Expt. No. 32. Test for Urea (Urease Test)

Principle: The urease enzyme decomposes urea to ammonia and carbon dioxide which together form ammonium carbonate, an alkaline substance, which changes the slightly acid reaction (yellow color) to alkaline reaction (pink color). Since urease is specific for urea only, positive test indicates the presence of urea.

$$CO(NH_2)_2 + H_2O \longrightarrow 2NH_3 + CO_2$$

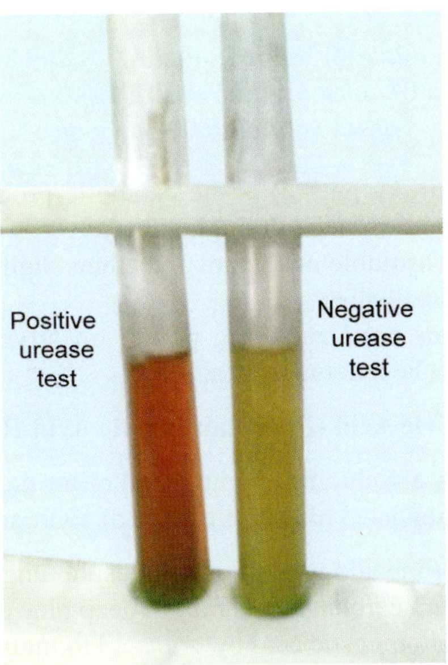

Procedure:-Take 2 ml of urine in one test tube and 2 ml of water in another. To each test tube add a drop of phenol red indicator and then add 2% sodium carbonate drop by drop to make the reaction in each tube alkaline (noted by appearance of pink color). Now add 1% acetic acid to both tubes drop by drop till the pink color just disappears (reaction changes to just acidic). Add a pinch of soyabean powder or Horse gram (which contains urease enzyme) to both the test tubes. Within a few minutes pink color reappears in the tube containing urine but not the other containing water.

Expt. No. 33. Hypobromite Test

Principle: Urea is decomposed by sodium hypobromite forming nitrogen gas.

$$CO(NH_2)_2 + 3NAOBr + 2NaOH \longrightarrow NaBr + N_2 + Na_2CO_3 + 3N_2O$$

Procedure: Take 5 ml of urine in a test tube. Add 1 ml of alkaline sodium hypobromite. Mix gently. A marked effervescence occurs due to evolution of nitrogen from urea.

Clinical Interpretation

Normal level: 25 – 30 Gm/day

Increased urinary urea:

- High protein diet
- Hematemesis
- Excess tissue breakdown as in high fever and severe wasting diseases.

Decreased urinary urea:

- Renal failure
- Severe hepatic insufficiency

- Low protein diet
- Severe acidosis

Uric Acid

Uric acid (2, 6, 8 trihydroxy purine) is the catabolic product of purines. It is synthesized in the liver and excreted through kidney.

Uric acid crystals are insoluble in alcohol and ether, slightly soluble in boiling water and quite soluble in alkalies.

The alkaline solutions have reducing power on silver and copper salts, phosphomolybdates and phosphotungstates.

Expt. No. 34. Test for Uric Acid (Phosphotungstic Acid Reduction Test)

Principle:-Uric acid is a reducing agent in alkaline conditions. It reduces phosphotungstic acid to tungsten blue in presence of sodium carbonate.

Procedure:-To 2 ml of urine add few drops of phosphotungstic acid reagent and a few drops of 20% sodium carbonate and mix. A deep blue color is produced due to the reduction of phosphotungstic acid by uric acid to tungsten blue.

Clinical Interpretation

Normal levels: 250 – 750 mgs/Day

Increased urinary uric acid:

- High purine diet (meat, legumes)
- Attacks of gout
- Leukemia
- Administration of cortisone or ACTH

Decreased urinary uric acid:

- Chronic renal failure.

Creatinine

Creatine is a normal constituent of muscle, where it is present as creatine phosphate. Under physiological conditions of temperature and pH, creatine undergoes spontaneous dehydration associated with ring closure to form creatinine.

Creatinine is normally excreted in urine. Normally, the daily output of creatinine is remarkably constant in an individual and depends on the muscle mass and shows little response to dietary change.

Expt. No. 35. Jaffe's Test (Test for Creatinine)

Principle: Creatinine in alkaline medium reacts with picric acid to form creatinine picrate which is reddish orange in color.

Procedure: To 3 ml of saturated picric acid solution, add 0.5 ml of 10% NaOH and divide it equally into two test tubes. To one test tube, add equal amount of urine and to the other same amount of water.

The tube containing urine assumes a deeper reddish orange color due to the formation of creatinine picrate whereas the control tube, containing water shows no change in its color.

Clinical Interpretation

Normal levels: 1–2 gms/day being higher in males than females
Increased urinary creatinine:
- Fever
- Myasthenia gravis
- Muscular atrophy, Myositis
- Hyperthyroidism
- Starvation

Ammonia

Ammonium ions are a constituent of normal urine and are present as ammonium salts. They are produced by the kidney and are not derived from the diet directly.

There is an increased excretion following the administration of inorganic acids, hepatic diseases and as a result of acid poisoning.

Expt. No. 36. Test for ammonia

Principle: Ammonia is liberated when urine is boiled in alkaline conditions.

Procedure: To 5 ml of urine, add 2% Na_2CO_3 till the solution is alkaline to litmus. Boil the solution. Place a piece of moistened red litmus paper at the mouth of the test tube. Note the change in color to blue due to evolution of ammonia.

Clinical Interpretation

Normal levels: 0.4 –1.0 gms/Day
Increased urinary ammonia:
- Acidosis
- Severe diabetes mellitus
- Starvation
- Delayed chloroform poisoning.
- Hepatic diseases

Decreased urinary ammonia:
- Alkalosis
- Damaged distal renal tubules as in renal failure
- Glomerulonephritis
- Addison's disease.

INORGANIC CONSTITUENTS OF URINE

The chlorides, sulfates and phosphates or sodium, potassium, calcium, and magnesium are the chief inorganic constituents of urine.

Sodium and potassium are found to the extent of 5 gm and 3 gm respectively in terms of their oxides Na_2O and K_2O in 24 hours urine.

Magnesium is also excreted in small quantity (0.05 to 0.2 gm per day).

Chlorides

Chlorides form the **chief anion of the urine**. The amount excreted in urine depends upon food and on the quantity of its loss through sweat. About 10 – 12 gm of chloride (as NaCl) is excreted daily.

Expt. No. 37. Test for Chloride

Principle: Chloride present in urine is precipitated as silver chloride with silver nitrate in the presence of nitric acid.

Procedure: Acidify 2 ml of urine with 2 drops of conc. HNO_3 and add to it 1 ml of 3% silver nitrate solution. A white precipitate of silver chloride is formed.

Clinical Interpretation

Normal levels: 10 –12 gm/day

Increased urinary chloride excretion:

- Addison's disease
- Polyuria of advanced chronic nephritis and diabetes

Decreased urinary chloride excretion:

- Excessive sweating
- Diarrhea and vomiting
- Diabetes insipidus
- Cushing's syndrome
- Extensive burns
- Pneumonia

Phosphates

Phosphates exist in urine as salts of sodium, potassium, ammonium, calcium and magnesium. In alkaline urine, phosphates crystallize out in characteristic shapes.

Expt. No. 38. Test for Inorganic Phosphorus

Principle: Upon warming with ammonium molybdate in the presence of conc. HNO_3, inorganic phosphate is precipitated as canary yellow ammonium phosphomolybdate.

Procedure: To 5 ml of urine, add a few drops of conc. HNO_3 and a pinch of ammonium molybdate. Warm it. Note the canary yellow color of precipitate or solution.

Clinical Interpretation

Normal levels: 0.8 –1.2 gm/day

Increased phosphates excretion:
- Hyperparathyroidism
- Administration of parathyroid hormone

Decreased phosphates excrtion:
- Hypoparathyroidism
- Rickets
- Diarrhea
- Acute infections and Nephritis – due to impaired kidney function.
- Pregnancy

Calcium

The amount of calcium excreted in urine is low (0.1 to 0.2 gm per day). Its excretion in urine increases during disturbed bone metabolism such as hyperparathyroidism.

Expt. No. 39. Test for Calcium

Principle: With potassium oxalate in acidic conditions calcium is precipitated as calcium oxalate.

Procedure: To 2 ml of urine, add 5 drops of 1% acetic acid and 5 ml of 2% potassium oxalate. White precipitate of calcium oxalate is formed.

Clinical Interpretation

Normal levels: 0.1 – 0.3 gm/day

Increased urinary calcium:
- Hyperparathyroidism
- Hyperthyroidism
- Hypervitaminosis D
- Multiple myeloma

Sulfates

Sulfates of urine come mainly from the oxidation of sulfur containing substances, e.g. protein.

It is present in urine in 3 forms:

a. Neutral sulfur (unoxidized) such as cystine, methionine, etc.
b. Oxidized sulfur – as sulfates of sodium, potassium, etc.
c. Ethereal sulfate – some compounds are detoxicated by conjugation with sulfate group which are known as ethereal sulfates (indoxyl sulfate, skatoxyl sulfate, etc). A small amount of sulfur is excreted in this form.

Expt. No. 40. Test for Sulfates

Principle: Sulfate is precipitated as barium sulfate with barium chloride.

Procedure: To about 3 ml of urine, add a few drops of conc. HCl and 1 ml of 10% barium chloride solution. A white precipitate of barium sulfate is produced. Presence of HCl prevents precipitation of phosphates.

Clinical Interpretation: Inorganic sulfates

Normal level: 0.7 – 1 gm/day

Increased:

- High protein diet

Decreased:

- Renal dysfunction.

Organic sulfates:

Normal level: 0.06 – 0.12 gms/day

Increased:

- Cystinuria
- Homocystinuria
- Cyanide poisoning

Recap of Normal Inorganic Constituents of Urine

Constituents	*Normal excretion*	*Clinical significance*
Sodium	100 – 200 mEq/24 hr	Increase: Acute tubular necrosis, adrenal insufficiency, diuretic therapy. Decrease: Prerenal failure, hyperaldosteronism.
Potassium	20 – 100 mEq/24 hr	Increase: Renal tubular acidosis, hyperaldosteronism, loop diuretic therapy Decrease: Renal failure, adrenal insufficiency.
Chloride	100 – 250 mEq/24 hr	Increase: Fasting, nephritis. Decrease: Addison's disease, metabolic acidosis.
Calcium	0.1 – 0.3 gm/24 hr	Increase: Vitamin D deficiency, hypoparathyroidism. Decrease: Hypervitaminosis D, Hyperparathyroidism.
Phosphorus	0.3 – 0.6 gm/24 hr	Increase: Metabolic acidosis, renal tubular defect, Hyperparathyroidism, vitamin D deficiency. Decrease: hypoparathyroidism, severe diarrhea.

Organic Constituents

Constituents	**Normal Excretion**	**Clinical significance**
Urea	18 – 35 gm/24 hr	Increase: High protein diet, fever. Decrease: Renal failure.
Creatinine	0.8 – 1.2 gm/24 hr	Increase: Muscular dystrophy. Decrease: Renal failure.
Uric Acid	0.3 – 0.8 gm/24 hr	Increase: High meat diet, high legume diet, leukemia, polycythemia, Von Gierke's disease Decrease: Renal failure.

Recap of Major Minerals in Urine

Mineral	*Daily Requirement*	*Dietary Source*	*Function*
Sodium	5 – 10 gm/day	Table salt, sea food, cheese, milk, egg	Chief cation of ECF, acid-base balance, maintaining osmotic pressure
Potassium	3 – 4 gm/day	Banana, dates, fruits, coconut, oats	Chief cation of ICF, neuromuscular excitation
Calcium	800 – 1500 mg/day	Milk & milk products, egg, ragi, cereals, leafy vegetables, beans	Formation of bones & teeth, Heart & muscle contraction, blood clotting, second messenger, nerve transmission
Phosohorus	500 – 1200 mg/day	Milk and milk products, cereals, fish, dates, leafy vegetables	Formation of bones & teeth, high energy as ATP, GTP, UTP, Nucleotides coenzymes as NADP
Sulfur	-	Sulfur containing amino acids, meat, beans, vegetables	Sulfur containing amino acids (methionine, cysteine), S containing vitamins (thiamine, biotin), heparin, chondroitin sulfate
Chloride	5 – 10 gm/day	Table salt, tomatoes, olives	Regulation of acid-base balance, HCl secretion in stomach

RELEVANT QUESTIONS–NORMAL CONSTITUENTS OF URINE

1. Name the normal constituents of urine.
2. What is the volume of urine output per day?
3. What is the normal color of urine?
4. What is the normal specific gravity of urine?
5. What is the pH of normal urine?
6. What is the chief anion of urine?
7. What is the normal level of calcium excreted in urine per day?
8. In which diseases calcium will be excreted in large quantities in urine?
9. What is the normal phosphate level in urine? In which disease the level will be increased?
10. How much ammonia is excreted per day?
11. Name the NPN substances present in urine.
12. How much of urea is excreted in urine per day ?
13. What is urea? What is the normal urea level in blood and in 24 hours urine?
14. How and where is urea formed in the body?
15. a. What is uric acid ?
    b. How much quantity of uric acid is excreted in urine per day?
16. What is creatinine? How is it synthesized?
17. What is the difference between creatine and creatinine?
18. What is normal excretory level of creatinine in urine?
19. What is oliguria, polyuria and anuria?
20. Name the tests to detect urea in urine. Which is better and why?
21. What is the test for determining uric acid in urine?
22. How is uric acid formed in the body?
23. What is the normal value of uric acid in urine and in blood?
24. What are the functions of phosphorus in the body?
25. What is the dietary source of chloride?
26. What are the functions of calcium in the body?

ABNORMAL CONSTITUENTS OF URINE

Substances which are not present in easily detectable quantities in urine of healthy individuals but are known to occur in urine under certain pathological conditions are called **Abnormal or "Pathological" constituents of urine.**

Before proceeding to detect the presence of abnormal constituents, note the physical characteristics (color, appearance, sedimentation and specific gravity) and chemical reaction with litmus paper.

The abnormal constituents which are routinely looked for in urine are reducing sugars, acetone bodies, protein, blood pigments, bile pigments, bile salts and urobilinogen.

REDUCING SUGARS

Even **normal urine** contains less than **100 mg of glucose** and **glucoronides in 24 hrs urine samples** but their amount is too small to cause reduction in Benedict's test.

A sample of urine which reduces Benedict's reagent under the conditions laid down for the test may be looked upon as containing reducing sugar unless otherwise proved. A positive Benedict's test does not necessarily indicate the presence of glucose in urine. However, positive Benedict's test is usually taken for glucose, unless other conditions are suspected and proved. If mucin is suspected, Benedict's test should be repeated after removing mucin with kaolin.

If the urine is from a pregnant or **nursing woman,** presence of **lactose** may be suspected.

Identification of different **reducing sugars (glucose, lactose, galactose, and fructose)** may be established by performing Seliwanoff's test, Osazones test, Fermentation test or Chromatography.

Expt. No. 41. Benedict's Reduction Test

Principle: Reducing sugars in urine under hot alkaline conditions tautomerize and form enediols which are powerful reducing agents. They reduce the cupric ions of Benedict's reagent to red cuprous oxide. The cupric hydroxide formed during the reaction is kept in solution by metal chelators like citrate.

Procedure: To 5 ml of Benedict's reagent in a test tube, add 0.5 ml (about 8 drops) of the urine. Hold the test tube firmly with a test tube holder and boil the contents for two minutes over a steady small flame. A turbidity or precipitate of green, yellow or red color indicates the presence of reducing substance in the urine (usually glucose). The various colors of the precipitate depend upon the concentration of sugar in urine and therefore a rough estimate of sugar in urine can be made in the following way. This is a **semiquantitative test.**

Color	Sugar Content (Grams %)
Blue color	Nil
Green precipitate	0.1 to 0.5
Yellow precipitate	0.6 to 1.0
Orange precipitate	1 to 1.5
Red precipitate	1.5–2.0

NOTE: Thymol, formaldehyde, Chloroform, Lactic acid, Vitamin C, and Dextrin give false positive test with Benedict's reagent.

BENEDICT SEMIQUANTITATIVE TEST

Acetone Bodies

When glucose metabolism is slow (as in diabetes mellitus) or when carbohydrate is not available (as in starvation), fat is excessively catabolized. As a result, acetoacetic acid and its derivatives, α hydroxy butyric acid and acetone, accumulate and are excreted in urine and expired through lungs respectively.

This condition is called "Ketosis" and these three compounds are together called acetone bodies or ketone bodies. The excretion of ketone bodies in urine is called "Ketonuria".

Expt. No. 42. Rothera's Test for Acetone Bodies

Principle: Acetone and acetoacetic acid form a purple colored complex with sodium nitroprusside in presence of ammonia.

The permanganate color reaction is presumably due to the formation of Ferro Penta Cyanide with the Isonitro compound of the ketone or the formation of such component with the isonitrosamine derivative of the ketone.

B-hydroxy butyrate does not respond to this test.

Procedure: Saturate 1 ml urine with solid ammonium Sulfate by shaking it, until some remains undissolved. Add 2 drops of freshly prepared solution of 5% Sodium Nitroprusside (or a small crystal of Sodium Nitroprusside and shake to dissolve in the urine) and mix.

Then, gently layer 2 ml of ammonia over the urine. At the junction of urine and ammonia layers a purple color ring appears if acetone bodies are present.

Note: The test cannot be regarded as negative until the mixture has stood for 10 minutes without developing this color.

Expt. No. 43. Gerhardt's Test (Acetoacetic Acid)

Principle: Acetoacetate reacts with ferric chloride to form a red-brown precipitate.

Procedure: To 5 ml of urine in a test tube, add 5% ferric chloride solution, drop by drop, till no precipitate of ferric phosphate is formed. Filter and to the filtrate add some more ferric chloride solution. A portwine color indicates the presence of acetoacetate.

Note: A similar color is given by large number of substances, such as aspirin, antipyrin, salicylates, etc. If the urine is boiled, acetoacetic acid is converted into acetone, but the other substances are not destroyed. A urine which gives a positive test before boiling and negative after boiling indicates the presence of acetoacetic acid.

Clinical Interpretation

Normal levels: Less than 1 mg/day
Increased ketone bodies excretion:
- Starvation
- Diet rich in fat but restricted proteins and carbohydrates
- Diabetic ketoacidosis
- Fevers and severe anemia
- Phosphorus poisoning
- Recurrent vomiting in children.

BLOOD PIGMENTS

Blood pigments, hemoglobin, present in the urine may be either in the form of **intact cells, hematuria,** or **free in solution, hemoglobinuria.**

In certain pathological conditions blood appears in urine. The presence of blood can be recognized by reddish color of urine and by microscopic examination of red blood cells, if it has not been hemolyzed.

Chemically, hemoglobin or its derivatives can be demonstrated by Benzidine test.

Expt. No. 44. Benzidine Test

Principle: Heme of hemoglobin decomposes hydrogen peroxide to nascent oxygen which oxidizes benzidine to bluish green colored product.

Procedure: In a dry test tube, dissolve a pinch of benzidine in about 1 ml of glacial acetic acid and add 1 ml of hydrogen peroxide to it. Then add 5 – 10 drops of urine to the test tube. The benzidine solution will turn deep blue, if blood pigment is present in the urine.

The test is very sensitive, as any test tube containing traces of uncleaned blood will give positive test. It should be noted that the blue color of benzidine produced by blood pigment is temporary and changes to brown within a few minutes. Repeat the test taking water as control.

Clinical Interpretation

Conditions of hematuria:

- Injury to urinary tract or kidney
- Urinary tract infection
- Benign or malignant tumors of kidney or urinary tract
- Urinary calculus
- Ruptured venous plexus of enlarged prostate

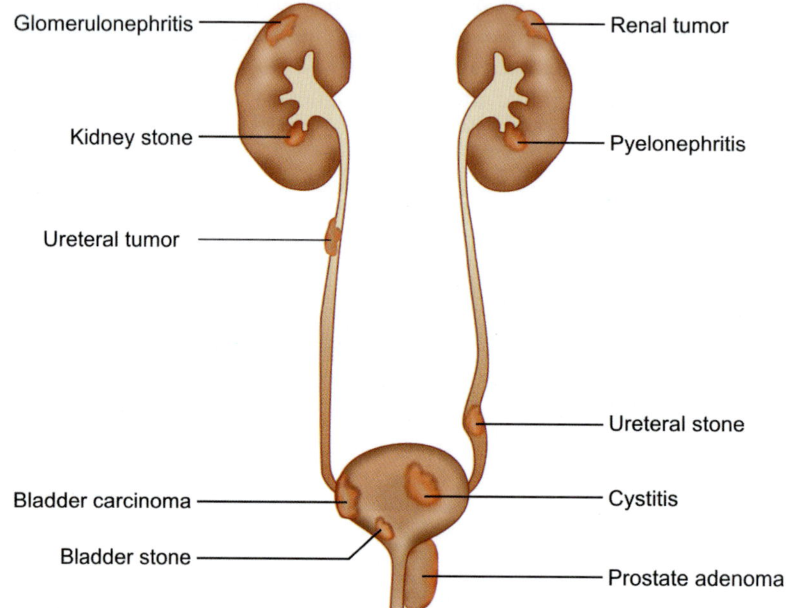

Common Renal and Postrenal Causes of Hematuria

Expt. No. 45. Guaiac Test

Principle: The guaiac test is based on the pseudo-peroxidase property of hemoglobin.

Procedure: Dispense 5 ml of fresh urine into a heat-resistant glass tube. Add freshly made gum guaiac reagent until turbidity forms. Add about 10 drops of freshly prepared hydrogen peroxide. Look for a blue color. If a blue color develops, boil the urine in a container of boiling water for 15 – 20 minutes.

Results

Blue color remains after boiling	Positive test for hemoglobin
Blue color disappears after boiling	Negative test.

Conditions of hemoglobinuria:

- Incompatible blood transfusion
- Hemolytic jaundice
- Severe burns
- Black water fever of falciparum malaria
- Paroxysmal hemoglobinuria due to exposure to severe cold
- Snake bites that cause acute hemolysis.
- Sickle cell disease crisis with severe hemolysis.
- March hemoglobinuria due to severe muscular exertion

HEMOGLOBINURIA:
The presence of free hemoglobin in urine is called Hemoglobinuria. It occurs with severe intravascular hemolysis when the amount of hemoglobin being released into the plasma is more than can be taken up by haptoglobin (the plasma protein that binds free hemoglobin to prevent it being lost from the body). The renal threshold for free hemoglobin is 1.0 – 1.4 g/l.

PROTEIN

The presence of heat coagulable protein in urine is called "Proteinuria". In such urine, generally **both albumins and globulins** are present.

Sometimes Bence Jones Protein (an immunoglobulin) or proteoses may be present. **Bence Jones protein** appears in urine in cases of **Multiple Myeloma.** It coagulates between 40 – 60°C, redissolves on further heating and coagulates on cooling to below 60°C. Of the various tests for protein in urine, the heat coagulation test is most commonly employed. The sulphosalicylic acid test is strongly recommended for routine work. A positive test should be checked by the heat coagulation test.

Clinical Interpretation

Conditions of Proteinuria:

- Nephrosis /Nephritis
- Tuberculosis
- Neoplasm
- Diabetic nephropathy
- Bence-Jones proteins in multiple myeloma

- Congestive heart failure
- Hypertension
- Strenuous exercise
- Pregnancy

Expt. No. 46. Sulphosalicylic Acid Test

Principle: Proteins are amphoteric in nature, i.e. they behave as acids in an alkaline medium and as bases in an acidic medium. In the presence of alkaloid reagents, like sulphosalicylic acid, they act as bases, and react with the acid to form an insoluble salt of protein sulphosalicylate.

Procedure: To 2 ml of clear urine (filter, if not clear) add a few drops of 20% sulphosalicylic acid. A turbidity of protein sulphosalicylate indicates the presence of a protein.

This test is positive for any protein. Hence presence of coagulable protein in urine must be confirmed by heat coagulation test.

Expt. No. 47. Heat Coagulation Test

Principle: When protein in urine is heated, its physical, chemical and biological properties are changed due to breaking up of certain bonds and the resultant change in the conformation of its molecules. This process is known as denaturation. However, when the coagulable proteins are heated at their isoelectric pH, a series of changes occur involving dissociation of the protein subunits (disruption of quaternary structure), uncoiling of the polypeptide chains (disruption of tertiary and secondary structure) and matting together of the uncoiled polypeptide chains (coagulation). While a denatured protein may be restored to its original structure and function by certain manipulations, coagulation is an irreversible process.

Procedure: Fill a test tube 3/4th full of clear urine. If the urine is turbid, filter it before performing the test. Heat the upper one-third of the test tube over a small flame. Turbidity is observed in the heated portion of the urine, well contrasted with the unheated lower portion, if coagulable protein or phosphate is present.

The presence of protein or phosphate can be distinguished by adding a drop of 1% acetic acid to the urine in which the phosphate dissolves but not the protein.

Even if turbidity does not appear on heating add 1 – 2 drops of acetic acid to the hot urine. Sometimes turbidity appears as addition of acetic acid brings the pH to its isoelectric pH.

BILE SALTS, BILE PIGMENTS AND UROBILINOGEN

The constituents or derivatives of bile that may appear in the urine are the bile pigments bilirubin and biliverdin, bile acids, chiefly glycocholic acid, urobilin and urobilinogen. Increased amounts of urobilin point to functional incapacity of the liver. Choluria is also present in obstructive jaundice. In hemolytic jaundice there is acholuria, though the urinary urobilinogen is raised.

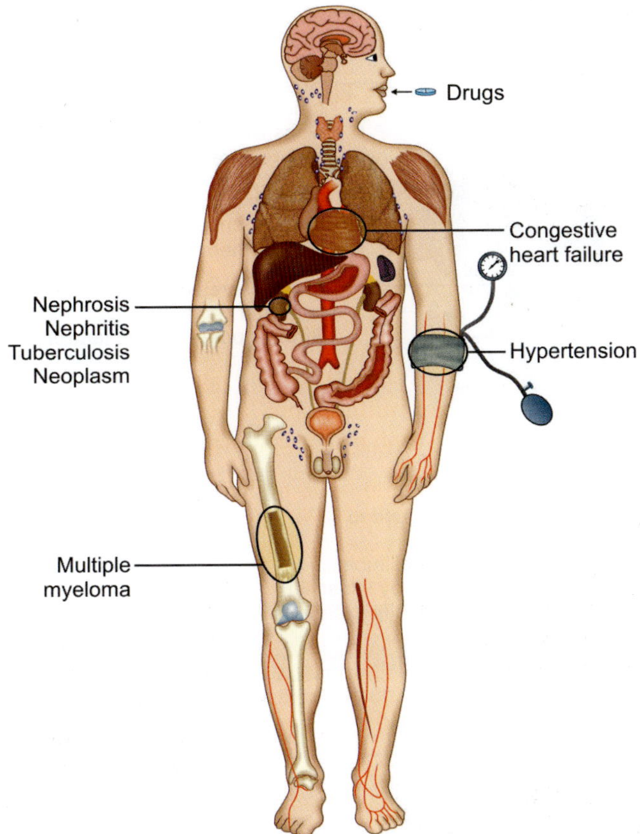

Common Causes of Proteinuria

BILE SALTS

Bile salts are the sodium taurocholate and sodium glycocholate.

Expt. No. 48. Hay's Test for Bile Salts

Principle:-Bile salts lower the surface tension of urine and allow the sulfur particles to sink.

Procedure: Fill 2/3rd of the test tube with urine and sprinkle a little sulfur powder over the urine. Note if sulfur particles sink which indicates the presence of bile salts. Repeat with water as control.

BILE PIGMENTS

The presence of bile pigments and acids together is associated with obstruction to the outflow of bile from the liver. This may be either intra or extra hepatic. Urinary excretion of bilirubin alone is seen in conditions of excessive hemolysis. In hepatocellular jaundice (infectious hepatitis) there are bile pigments in urine even before jaundice becomes clinically detectable.

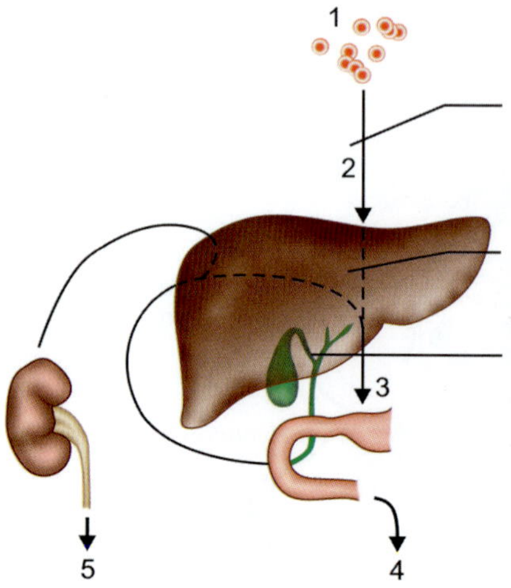

RBC-hemolysis releases bile pigments (unconjugated bilirubin) into circulation which binds with albumin and transported to liver.
Site for prehepatic jaundice

Bilirubin is conjugated with glucoronic aoid in the liver and secreted in the bile.
Site for hepatic jaundice

Bile is secreted into the intestine where the intestinal bacteria convert it to urobilinogen.
Site for post-hepatic jaundice

1. RBC lysis release bilirubin
2. Bilirubin transported in blood bound to albumin. (unconjugated)
3. Released in bile as conjugated bilirubin.
4. Conjugated bilirubin acted upon by intestinal bacteria form urobilinogen, which is excreted in feces as stercobilinogen.
5. Excreted in urine as urobilinogen.

Bilirubinuria: Bilirubin is not normally detected in the urine. When it is found, the condition is referred to as bilirubinuria. Urine containing 8.4 μmol/l (0.5 mg%) or more of bilirubin has a characteristic yellow-brown color (hepatocellular jaundice) or a yellow-green appearanace (obstructive jaundice)

Expt. No. 49. Fouchet's Test for Bile Pigment

Principle: Barium chloride reacts with sulfate radicals in urine to form barium sulfate. If bilirubin is present in urine, it adheres to the precipitate and is detected by oxidation of bilirubin (yellow) to biliverdin (green) with ferric chloride in the presence of trichloroacetic acid present in Fouchets reagent

Procedure: Take 5 ml of urine in a test tube, add 1 – 2 drops of saturated magnesium sulfate solution and about 5 ml of 10% $BaCl_2$ to it. Mix the contents and let the tube stand for a while for the precipitate to settle down. Decant the supernatant and filter the rest. After the filtration is complete spread the filter paper over the work bench, and add a drop of Fouchet's reagent to one part of the precipitate on the filter paper. Bile pigment is indicated if blue or greenish blue color appears at the place where Fouchet's reagent has been added.

Note: Freshly passed urine is required. It should be protected from daylight and fluorescent light because bilirubin is rapidly oxidized by ultraviolet light to biliverdin which is not detected by the reagents used in the Fouchet's test, Ictotest tablet test, or bilirubin strip tests.

UROBILINOGEN

Urobilinogen normally is present in the urine in amounts sufficient to give a positive test in dilutions of 1 : 10 to 1 : 20. Its absence indicates complete biliary obstruction while increased amounts are associated with excessive blood destruction.

Expt. No. 50. Ehrlich's Test for Urobilinogen

Urobilinogen is formed from bile pigments in intestine and is absorbed to be excreted in urine. Urobilinogen is excreted in small amount in normal urine but is excreted in greater quantity when large amount of bile pigments reach intestine as in hemolytic jaundice. Absence of urobilinogen from urine indicates obstruction of biliary passages.

Principle: Urobilinogen reacts with Ehrlich's reagent to form red color which intensified on addition of sodium acetate.

Procedure: To 2 ml of urine, add about 2 ml Ehrlich's aldehyde reagent. Mix and allow it to stand for 10 minutes. Add sodium acetate powder till saturation. Normal urine gives out a pink color but a distinctly red color suggests the presence of increased amount of urobilinogen. This test should be performed on fresh urine only.

Note: Bilirubin interferes with the reaction and therefore if present it must first be removed by reacting the urine with barium chloride **(Watson's Modification).**

Method – Take two test tubes and label one 'T' (test) and other 'C' (control). Dispense 5 ml of fresh urine into each tube. If the urine contains bilirubin, mix equal volumes of the specimen with 0.48 mol/l barium chloride. Centrifuge or filter to obtain clear supernatant or filtrate to test for urobilinogen. Add 0.5ml Ehrlich's reagent to tube "T" and mix. Add 0.5 ml of 50% v/v hydrochloride to tube 'C' and mix. Leave both tubes at room temperature (20 – 28°C) for 5 minutes. Looking down through the tubes, examine for a definite red color in the tube 'T' as compared with tube 'C'.

Results

Red color in tube T	Increased urobilinogen
Pink color in tube T	Normal urobilinogen
T & C appear similar	No urobilinogen

How to check that the red color is due to urobilinogen and not due to porphobilinogen*?

Add 1 – 2 ml of chloroform or amyl alcohol. Mix well and allow to settle. Look for red color in the chloroform or amyl alcohol layer, indicating the presence of urobilinogen. If red color is not in this layer it is due to porphobilinogen*.

*Porphobilinogen is excreted in the urine of patients with acute intermittent porphyria, a rare inherited disease that affects nerves and muscles.

Note: False reactions—A false negative reaction may occur if the urine contains nitrite as in some bacterial urinary infections. The nitrite will oxidize the urobilinogen to urobilin which is not detected by Ehrlich's reagent. A negative reaction may also occur if a patient is receiving intensive antimicrobial therapy. The antimicrobials will reduce the number of bacteria in the intestine and so prevent urobilinogen being formed.

Besides porphobilinogen, other substances which react with Ehrlich's reagent include the metabolite indican and drugs like p-amino salicylic acid and sulphonamides.

Point-of-care testing (POCT) is defined as diagnostic tests performed at or near the site of patient care to deliver the reports early for prompt treatment. Urine test strips (uristix) are used to detect reducing sugars, protein, blood, bilirubin, ketones, pH, and specific gravity in urine.

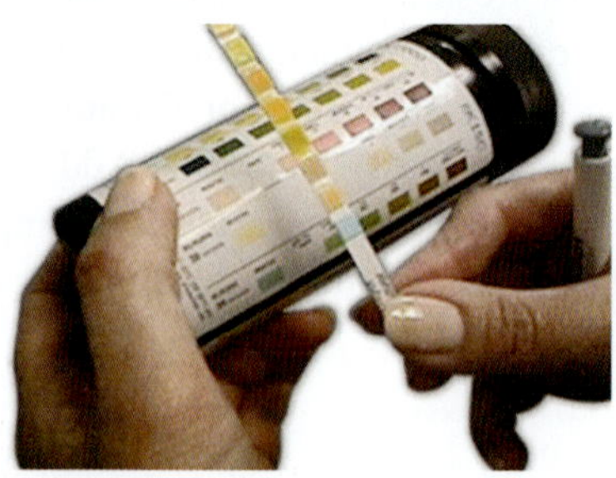

Recap of Abnormal Constituents of Urine

Abnormal Constituents	Clinical Conditions/ Causes	Detection Tests
	REDUCING SUGARS	
Glucose	Diabetes mellitus, Renal glysosuria, Alimentary glycosuria, thyrotoxicosis, Cushing's disease.	Benedict's test, Glucose oxidase test, Chromatography
Fructose	Hereditary fructose intolerance: Aldolase B deficiency Essential fructosuria: Fructokinase deficiency	Seliwanoff's test, Chromatography
Galactose	Galactosemia: Galactose 1 phosphate uridyl transferase deficiency	Mucic acid test, Chromatography
Xylulose	Essential pentosuria: Xylitol dehydrogenase deficiency	Bial's test, Chromatography
Lactose	Lactosuria: Lactation, last trimester of pregnancy	Lactosazone test, Chromatography
	PROTEINS	
ALBUMIN, if mild injury, IgG (non-selective) if severe injury.	Glomerular proteinuria: Nephrotic syndrome, Nephritis, diabetic nephropathy	Heat coagulation test, Sulphosalicylic acid test, Immunoturbidometry
Bence Jones proteins	Overflow proteinuria: Multiple myeloma	Bradshaw test, Heat coagulation test, Immunoelectrophoresis
β-2 microglobin	Tubular proteinuria: Acute tubular necrosis, renal transplant rejection	Nephelometry
	LIPIDS	
KETONE BODIES	Ketoacidosis: Diabetes mellitus, Starvation	Rothera's test
Triacylglycerols/cholesterol	Lipiduria: Nephrotic syndrome	
	BLOOD	
BLOOD	Glomerulonephritis, calculus, tumor, trauma	Benzidine test
	BILE PIGMENTS	
BILIRUBIN	Obstructive jaundice: Gallstones, carcinoma of head of pancreas, hepatitis	Fouchet's test
	BILE SALTS	
SODIUM/POTASSIUM SALTS OF GLYCOCHOLATE AND TAUROCHOLATE	Obstructive jaundice	Hay's test
	AMINO ACIDS	
PHENYLALANINE	Phenylketonuria: Phenylalanine hydroxylase deficiency	Ferric chloride test

RELEVANT QUESTIONS—ABNORMAL CONSTITUENTS OF URINE

1. What is glycosuria? In which conditions it occurs?
2. Name the test to identify reducing sugar in urine.
3. What is proteinuria? In which diseases protein will be present in urine?
4. Name the proteinuria in which light chains of Immunoglobulins are excreted in urine?
5. In which disease does this proteinuria occur?
6. What is the test for identifying Bence-Jones proteins?
7. What is ketosis?
8. In which diseases does ketonuria occurs?
9. Name the ketone bodies. How and where are they synthesized?
10. Which test shows the presence of ketone bodies in urine?
11. Why β hydroxyl butyrate does not respond to Rothera's test?
12. What is hematuria? Give some causes.
13. What is hemoglobinuria? In which diseases there will be hemoglobinuria?
14. Why do you add acetic acid in heat coagulation test?
15. How does one differentiate between hematuria and hemoglobinuria?
16. Name the condition in which bile salts and bile pigments are present in urine.
17. What is jaundice? What are its types?
18. What are bile salts? How are they derived in human body?
19. Name the bile pigments. From which substances are they derived?
20. What are the differences between urobilins and urobilinogens?
21. What are the urinary findings in hemolytic and obstructive jaundice?
22. In which conditions both glucose and ketone bodies are excreted in urine? How can one differentiate between the two disorders?

Chapter 3 Quantitative Analysis

INTRODUCTION TO QUANTITATIVE ANALYSIS

Biochemical investigations are routinely requested to support the clinical specialties for screening, diagnosis and monitoring of patients. Most biochemical tests are performed on blood or urine samples. However, occasionally analysis of body fluids like CSF, pleural and peritoneal fluids, feces and calculi are also performed.

Quantitative analysis gives an exact amount of the substance present in the sample unlike qualitative tests which only give information of the presence or absence of a compound in the given sample. Semi-quantitative tests (e.g. Benedict's test, Heat coagulation test) on the other hand give an idea or approximation on the range of substance present in the sample, like glucose in urine present between 0.5 – 1.0 gm/100 ml in Benedict's test.

Blood samples are collected in color coded vacutainers which are specific for the type of parameters advised for a patient. A list of the commonly used vacutainers is given in the table below.

Color Coded Vacutainers for Sample Collection

Vacutainer	*Color code*	*Sample*	*Tests performed*
	Red	Serum	Chemistry, HIV, Hepatitis, Blood drug screens, Therapeutic drug monitoring
	Yellow	Whole blood or plasma	DNA HLA
	Green	Sodium /lithium heparin	Plasma determination of ammonia
	Purple/ Lavender	Whole Blood (EDTA)	Hematology
	Light Blue	Plasma (Citrate) (Sodium Citrate)	Coagulation (e.g. PT and PTT)
	Gray	Whole Blood (Sodium fluoride/ Potassium oxalate)	Blood alcohol and plasma glucose

In order to perform biochemical tests, it is essential to collect proper specimen otherwise the test result would be meaningless, and at times, misleading. Therefore, a good specimen is one which is collected from the correct patient, at correct time, in the correct container and subsequently correct handling to be analyzed eventually.

Most clinical biochemistry tests are performed on whole blood, serum or plasma. Plasma is the fluid component of blood after centrifugation when clotting is prevented by addition of anticoagulants like heparin, EDTA, etc. When the blood is allowed to clot the fluid, part which separates after centrifugation is called serum.

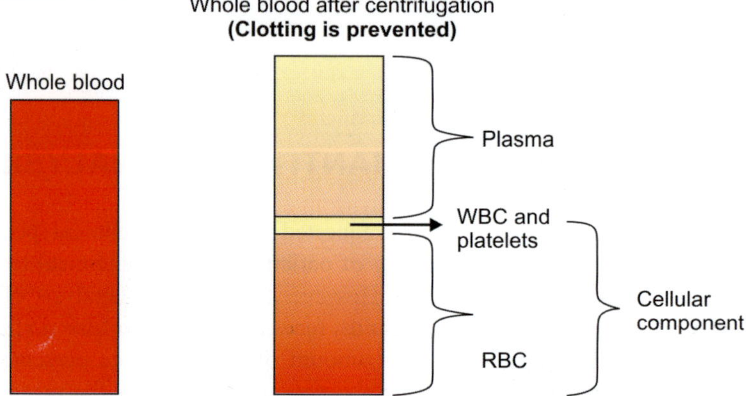

Fluid component of blood is called plasma which comprises 55% of the total blood volume. Plasma contains proteins, sugars, vitamins, minerals, lipoproteins and clotting factors. 95% of plasma is water.

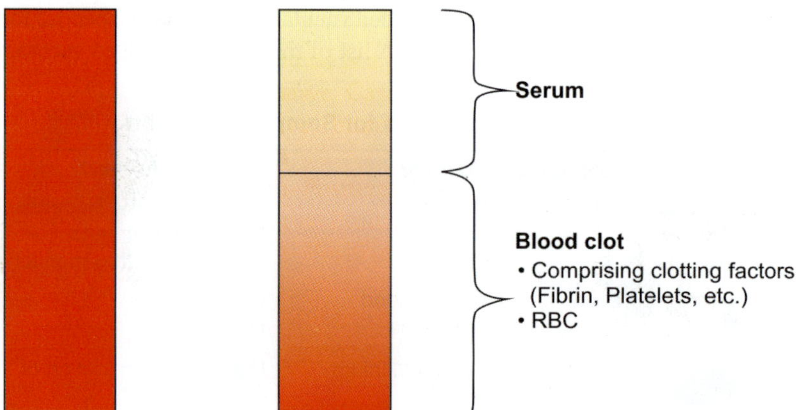

Blood is collected and allowed to clot on standing. There is formation of an insoluble fibrin clot. If the blood is centrifuged, the fluid part separates out and is called serum.

Finally, no matter how meticulously an analyst performs the test in a laboratory, analytical results are subjected to blunders which could be systemic or random type of errors. The analytical errors can be markedly reduced if one is aware of the type of errors that could occur in the entire analytical process. The errors are broadly classified into three types: Pre-analytical, Analytical and Post-analytical. The type of errors in each class is given below.

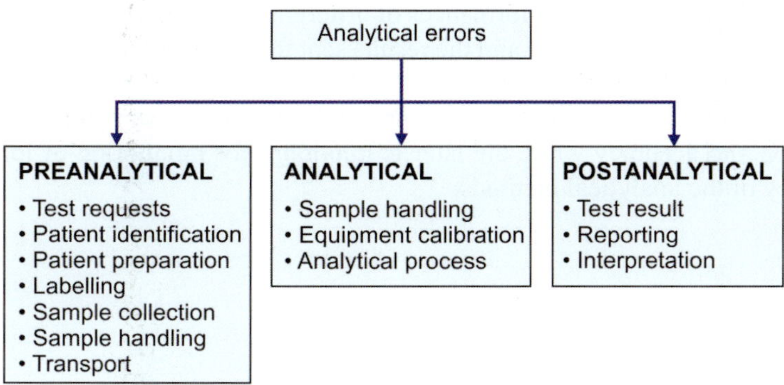

QUALITY CONTROL

To minimize or avoid the various analytical errors, laboratories need to follow quality control measures and ensure that the patients' results are reliable. Reliability of results indicate both accuracy (i.e. how close a result is to its true/actual results) and precision (i.e. how consistent the result is when performed at different times).

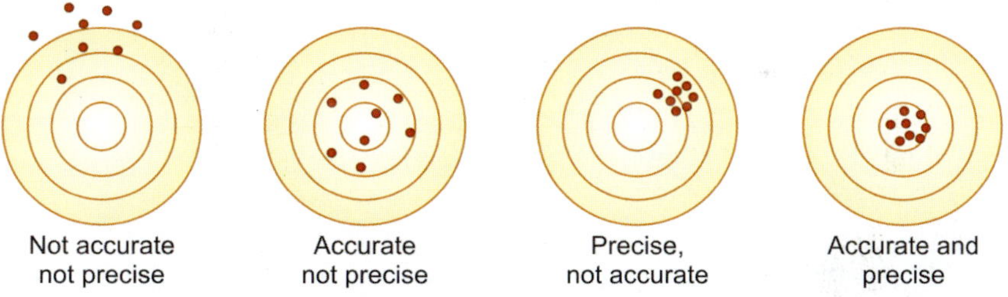

The quality control measures include:

- Checking water quality, calibration of glass wares and instruments that may alter the results.
- Performing routine maintenance of instruments.
- Regular monitoring of control charts (Levey-Jennings control chart)
- Participating in proficiency testing programs.

Quality control is of broadly of two types:

- Internal quality control **(IQC)**
- External quality control **(EQC)**

Internal quality control:

- IQC is run everyday or every time the test is performed. The control material is a representative of the test material in composition, state of physical preparation and concentration range of the analyte.IQC checks the correct execution of the procedure. Thus, a known concentration is taken and compared with the values obtained. These procedures are monitored in a single laboratory.

External quality control:

- EQC is comparing the performances of different laboratories. The procedure is done mostly once a month and the results sent to the concerned program center for monitoring and feedback.

IQC and EQC procedures are complementary wherein the former monitors the daily precision and accuracy while the later is important for monitoring of long-term accuracy of the analytical methods.

COLORIMETRY

Colorimetry is a technique which is frequently used in biochemical estimations. It involves the quantitative estimation of color. Thus, a substance to be estimated colorimetrically must be either colored or, most often, capable of forming chromogens through the addition of reagents. The instrument commonly used, known as colorimeter, is in fact absorptiometer. It measures the amount of light absorbed in these estimations.

The parts of a colorimeter include a source of **light** and a device for selecting light of narrow wavelength (Fig. 3.1). The common colorimeters have a set of **filters**. For example, a green filter absorbs all the component colors of white light except green light which is allowed to pass through. The light that is transmitted through the green filters has a wavelength from 500 to 560 nm. Similarly, other suitable filters can be used to select light of narrow wavelengths. In sophisticated instruments, a diffraction grating or prism is used in place of filters so that the transmitted light has a very narrow bandwidth.

The transmitted light passes through a compartment in which the colored solution is placed in a **cuvette or test tube**. The colored solution absorbs some light and the residual light falls on a **detector**. The detectors are photosensitive elements which convert the light into electrical signal. The electrical signal generated is directly proportional to the intensity of light falling on the detector. This signal is quantitated by a **galvanometer** and read as **optical density (OD) value.**

Colored solutions have the property of absorbing certain wavelengths of light and transmitting others. The color of the solution depends upon the transmitted light. For example, a solution of hemoglobin absorbs blue-green light and transmits the complementary red color and hence the solution appears red. In colorimetry, filters are chosen appropriately so that absorption is maximum at the selected wavelength band.

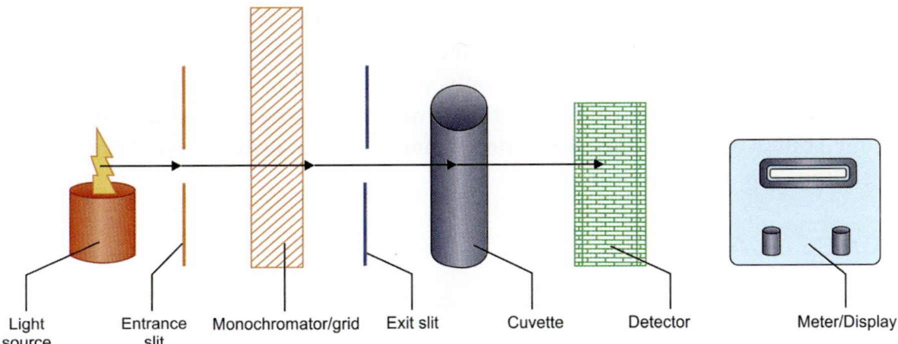

Fig. 3.1: Schematic diagram of a colorimeter

Complementary Colors Used in Colorimetry for Colored Solutions

Solution color	*Filter used*	*Peak transmission range (Nm)*
Bluish Green	Red	580 – 680
Blue	Yellow	520 – 580
Purple	Green	490 – 520
Red	Blue-green	470 – 490
Yellow	Blue	430 – 470
Yellowish-green	Violet	400 – 430

According to **Beer's law**, the **amount of light transmitted through a colored solution decreases exponentially with increase in concentration of the colored substance.**

In colorimetric estimation, it is necessary to prepare a **'reagent blank'**, **'test'** and a **'standard'**. The test solution is made by treating a specified volume of the blood filtrate or other specimen with reagents as mentioned in the procedure. A **standard solution** is prepared by similarly treating simultaneously, a solution of **pure substance whose concentration is known.** Since some color comes from the reagents specified in the procedure which is in addition to the color produced by the substance which is desired to be estimated, a blank is also run by similarly treating at the same time a same volume of distilled water.

STEPS IN THE OPERATION OF THE PHOTOELECTRIC COLORIMETER

Place the glass filter recommended in the procedure in the filter slot. Fill the colorimeter tube (cuvette) to about three fourth with the distilled water and place in cuvette slot.

Switch 'on' the instrument and allow it to warm up for 4 – 5 minutes.

Adjust to zero optical density. Take 'blank' solution in another tube and with this placed in the cuvette slot, read the optical density (B).

Take 'test' solution in the cuvette, and as with 'blank' read the O.D. (T). Finally take 'standard' solution in the cuvette and record O.D. (S).

Satisfactory results are obtained only when the O.D. values are in the range 0.1 – 0.7. **The reading should be taken with the lighter solution first (lower conc.) followed by the darker solutions (more conc.) in order to avoid erroneous readings due to carryover effect.**

CALCULATION

$$\text{Conc. of the test} = \frac{OD_T - OD_B}{OD_S - OD_B} \times \frac{\text{Conc. of std}}{\text{Vol. of test}} \times 100$$

$$= \frac{T-B}{S-B} \times \frac{\text{Conc. of std}}{\text{Volume of test}} \times 100$$

BLOOD GLUCOSE ESTIMATION

Normally, the body maintains the blood glucose level at a **reference range** between about 70 - 105 mg/dL in serum or plasma and 65 – 95 mg/dL in whole blood. The human body naturally tightly regulates blood glucose levels as a part of metabolic homeostasis.

Two major methods have been used to measure glucose. The first is a chemical method exploiting the ***nonspecific reducing*** property of glucose in a reaction with an indicator substance that changes color when reduced. Since other blood compounds also have reducing properties (e.g., urea, which can be abnormally high in uremic patients), this technique can produce erroneous readings in some situations (5 to 15 mg/dl has been reported). The more recent technique, ***using enzymes*** specific to glucose as substrate, is less susceptible to this kind of error. The two most common employed enzymes are Glucose Oxidase and Hexokinase. The enzymatic hexokinase method is the basis for the reference method for the determination of glucose in serum or plasma.

The international standard way of measuring blood glucose levels are in terms of a **molar concentration,** measured in mmol/L (millimoles per liter; or millimolar, abbreviated mM). In the United States, mass concentration is measured in mg/dL (milligrams per deciliter).

Since the molecular weight of glucose $C_6H_{12}O_6$ is about 180 g/mol, for the measurement of glucose, the difference between the two scales is a factor of 18, so that 1 mmol/L of glucose is equivalent to 18 mg/dL.

Collection of blood in clot tubes for serum chemistry analysis permits the metabolism of glucose in the sample by blood cells until separated by centrifugation. Red blood cells, for instance, do not require insulin for intake of glucose from the blood. Higher than normal amounts of white or red blood cell counts can lead to excessive glycolysis in the sample, with substantial reduction of glucose level if the sample is not processed quickly. Ambient temperature at which the blood sample is kept prior to centrifuging and separation of plasma/serum also affects glucose levels.

At refrigerator temperatures, glucose remains relatively stable for several hours in a blood sample. At room temperature (25°C), a loss of 7–10 mg/dL (or 0.4 mmol/L) of total glucose per hour should be expected in whole blood samples. Loss of glucose under these conditions can be prevented by using fluoride tubes (i.e., gray-top) since fluoride inhibits glycolysis. However, these should only be used when blood will be transported from one hospital laboratory to another for glucose measurement. Red-top serum separator tubes also preserve glucose in samples after being centrifuged isolating the serum from cells.

To prevent contamination of the sample with intravenous fluids, particular care should be given to drawing blood samples from the arm opposite the one in which an intravenous line is inserted.

The fasting blood glucose level, which is measured after a fast of 8 hours, is the most commonly used indication of overall glucose homeostasis, largely because disturbing events such as food intake are avoided. The metabolic response to a carbohydrate challenge is conveniently assessed by a postprandial glucose level drawn 2 hours after a meal or a glucose load. In addition, the glucose tolerance test, consisting of several timed measurements after a standardized amount of oral glucose intake, is used to aid in the diagnosis of diabetes.

Fasting Blood Glucose

Glucose level	*Indication*
From 70 to 99 mg/dL (3.9 to 5.5 mmol/L)	Normal fasting glucose
From 100 to 125 mg/dL (5.6 to 6.9 mmol/L) (pre-diabetes)	Impaired fasting glucose
126 mg/dL (7.0 mmol/L) and above on more than one testing occasion	Diabetes

Expt. No. 51. Blood Glucose Estimation
Method: Ortho-toluidine mono step method
Aim: To estimate blood glucose in the given sample
Reagents:
Reagents required:
1. Ortho-toluidine reagent
2. Glucose standard: 100 mg/dl in Saturated benzoic acid

Principle: Ortho-toluidine reacts quantitatively with aldehyde group of aldohexoses in presence of hot glacial acetic acid to form a glycosylamide and then a Schiff base which gives a bluish green colored solution which is recorded at 620 nm.

Procedure:

Mix thoroughly and place the tubes in the boiling water bath for exactly 10 minutes. By using tap water, cool the tubes to room temperature. Measure the O.D of test, standard and blank at 620 nm or red filter.

	Test	*Standard*	*Blank*
O-toluidine reagent	5 ml	5 ml	5 ml
Serum	0.05 ml	-	-
Glucose Standard (100 mg/dl)	-	0.05 ml	-
Distill Water	-	-	0.05 ml

Calculations:

$$\text{Serum glucose in mg/dl} = \frac{\text{OD of test}}{\text{OD of std}} \times \frac{\text{Eff. conc of std}}{\text{Eff. vol of test}} \times 100$$

$$= \frac{OD_T - OD_B}{OD_S - OD_B} \times \frac{0.05}{0.05} \times 100$$

$$= \frac{\text{OD of test}}{\text{OD of std}} \times 100 = \text{mg/dl}$$

Note: Hemolyzed, lipemic or icteric samples will interfere with this method.
Normal Range – Fasting blood glucose is 70–110 mg/dl

CLINICAL CONDITIONS

Hyperglycemia (Fig. 3.2) – Increase in blood glucose above normal.
- Diabetes mellitus

- Hyperthyroidism, hyperpituitarism and increased adrenocortical activity.

Hypoglycemia (Fig. 3.3) – Decrease in Blood glucose above normal.

- Overdose of insulin for diabetic treatment.
- Hypothyroidism, hypopituitarism and hypoadrenalism, insulinoma

Other Methods

1. Chemical methods based on (a) oxidation-reduction reactions with alkaline copper reduction and alkaline ferricyanide reduction and (b) condensation reactions
2. Enzymatic methods like glucose oxidase method and hexokinase method – The hexokinase is reference method.

GLUCOSE ESTIMATION

Glucose can be measured in the whole blood, serum or plasma, but plasma is recommended for diagnosis of diabetes mellitus. Furthermore, it is the WHO/ADA CRITERIA and not reference ranges that are used for the diagnosis of diabetes mellitus. Concentration of glucose is higher in arterial than in venous samples.

WHO/ADA criteria for diagnosis of diabetes mellitus.

Fasting : ≥126 mg/dl (7.0 mmol/l)

2-hr post-glucose load ≥ 200 mg/dl (11.1 mmol/l)

I. CHEMICAL METHODS

A. Oxidation-reduction reaction	
Glucose + Alkaline copper tartarate $\xrightarrow{\text{Reduction}}$ Cuprous oxide	
	1. Alkaline Copper Reduction
Folin-Wu method	Cu^{++} + Phosphomolybdic acid $\xrightarrow[\text{Phosphomolybdenum}]{\text{Oxidation}}$ Blue end-product
Benedict's method	• Modification of Folin-Wu method for qualitative urine glucose
Nelson-Somogyi method	Cu^{++} + Arsenomolybdic acid $\xrightarrow[\text{Arsenomolybdenum}]{\text{Oxidation}}$ Blue end-product
	2. Alkaline Ferricyanide Reduction
B. Condensation	
Ortho-toluidine method	• Uses aromatic amines and hot acetic acid • Forms glycosylamine and Schiff's base which is emerald green in color • This is the most specific method, but the reagent used is toxic.

II. ENZYMATIC METHODS

A. Glucose oxidase

$$\text{Glucose} + O_2 \xrightarrow[\text{Oxidation}]{\text{Glucose oxidase}} \text{Cuprous oxide}$$

Saifer–Gerstenfeld method	H_2O_2 + O-dianisidine $\xrightarrow[\text{Oxidation}]{\text{Peroxidase}}$ H_2O + oxidized chromogen	Inhibited by reducing substances like BUA,bilirubin, glutathione, ascorbic acid
Kodak Ektachem	• A dry chemistry method • Uses reflectance spectrophotometry to measure the intensity of color through a lower transparent film	
Glucometer	• Home monitoring blood glucose assay method • Uses a strip impregnated with a glucose oxidase reagent	

B. Hexokinase

$$\text{Glucose} + \text{ATP} \xrightarrow[\text{Phosphorylation}]{\text{Hexokinase + Mg ++}} \text{G} - 6PO_4 + \text{ADP}$$

$$\text{G-6}PO_4 + \text{NADP} \xrightarrow[\text{Oxidation}]{\text{G-6PD}} \text{G-Phosphogluconate} + \text{NADPH} + H^+$$

- NADP as cofactor
- NADPH (reduced product) is measured in 340 nm
- More specific than glucose oxidase method due to G-6PO_4, which inhibits interfering substances except when sample is hemolyzed

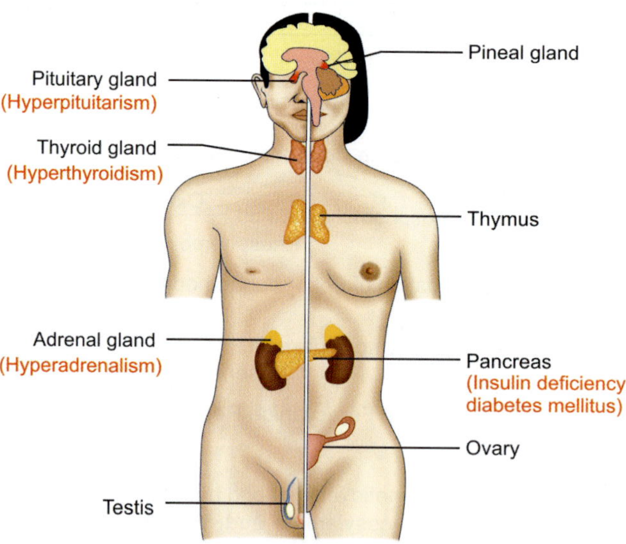

Fig. 3.2: Common causes of hyperglycemia

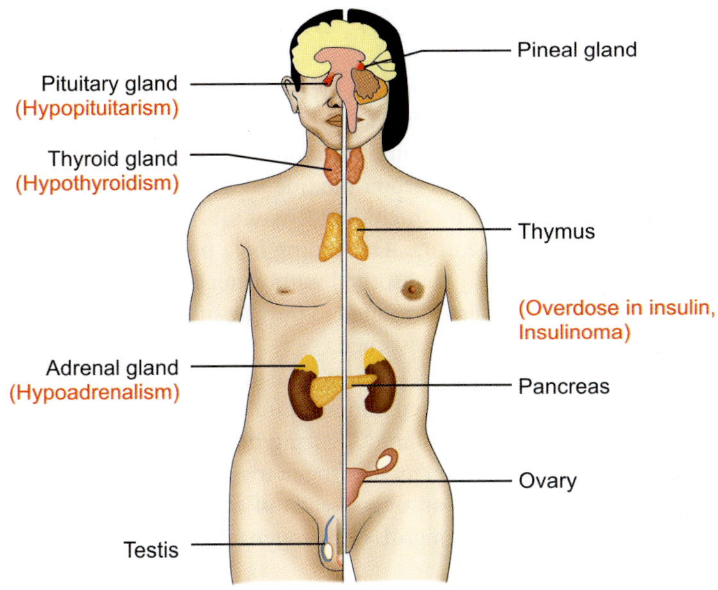

Fig. 3.3: Common cause of hypoglycemia

RELEVANT QUESTIONS

1. What is the normal fasting blood glucose level?
2. Name the conditions in which the blood glucose level is increased?
3. Name the conditions in which blood glucose level is decreased?
4. What are the methods by which blood glucose can be measured in the laboratory?
5. By which method have you estimated the test?
6. What is the principle of this procedure?
7. What is the importance of "blank" in quantitative analysis?
8. What is the importance of "standard" in quantitative analysis?
9. What is the color produced due to?
10. At what wave length you took the reading?
11. What is the color of the filter you used?
12. What is the composition of ortho-toluidine reagent?
13. What are the hormones regulating blood glucose level?
14. What are the tests for diagnosing diabetes mellitus?
15. What other sample is collected in addition to blood in a diabetes mellitus patient?
16. What tests are performed with the urine sample and why?

ESTIMATION OF BLOOD UREA

Urea is the main waste product of protein breakdown. It is formed in liver by the reactions of Krebs urea cycle. Amino acids are deaminated (nitrogenous amine group is removed), releasing ammonia. The ammonia, which is toxic to the body, is detoxified by combining with carbon dioxide to form urea which passes into the circulation and is excreted by the kidneys. Not all the urea that is filtered by the glomeruli is excreted in the urine. Depending on a person's state of hydration, 40–70% of the urea is passively reabsorbed with water and returned to the blood. The rate of reabsorption is inversely related to the rate of urine flow. When the rate of urine flow is low, more urea is absorbed.

The **blood urea nitrogen** (BUN) test is a measure of the amount of nitrogen in the blood in the form of urea. Normal human adult blood should contain between 7 to 21 mg of urea nitrogen per 100 ml (7–21 mg/dL) of blood.

To convert from mg/dL of blood urea nitrogen to mmol/L of **urea**, divide by 2.8 (each molecule of urea having 2 nitrogens, each of molar mass 14 g/mol).

Urea (in mmol/L) = BUN (in mg/dL of nitrogen) / 2.8

BLOOD UREA ESTIMATION

As a test of renal function, the measurement of plasma urea is less useful than the measurement of plasma creatinine. A number of extra renal factors influence the circulating urea concentration, like state of hydration and dietary intake, increased protein catabolism, reabsorption of blood proteins after gastrointestinal hemorrhage and treatment with cortisol or its synthetic analogues.

To convert BUN to urea in mg/dL by using following formula:

Urea= BUN × 2.14;

MW of urea = 60; urea nitrogen = 28;

Therefore, BUN = 60/28

Expt. No. 52. Estimation of Blood Urea

Method: Diacetyl-monoxime method

Aim: To estimate blood urea in the given sample

Principle: Under acidic conditions when urea is heated with compounds containing two adjacent carbonyl groups, such as diacetyl ($CH_3COC{=}NOHCH_3$) in presence of ferric ions and thiosemicarbazide pink colored diazine is formed at 520 nm.

REAGENTS

1. Sodium tungstate 10%
2. Sulfuric acid 2/3 N
3. Diacetyl monoxime 2% solution (Reagent is stable at room temperature for 1 year)

4. Thiosemicarbazide – 40% solution (Reagent is stable at room temperature for 6 months)
5. Sulfuric acid-phosphoric acid reagent (Reagent is stable at room temperature for 1 year)
6. Stock standard solution of urea 250 mg/100 ml
7. Working standards 1 in 100 dilution
8. Working reagent: Mix one part each of reagent 3 and 4 and two parts of reagent 5. Prepare fresh each day.

Preparation of protein free filtrates (PFF)

DW	3.5 ml
Blood	0.1 ml
10% Sod. Tungstate	0.2 ml
2/3 NH_2SO_4	0.2 ml
	----------
	4 ml

(1 ml of protein free filtrate = 0.025 ml of blood)

Urea solution is given intravenously for Sickle cell anemia. Urea prevents the Roulex formation and prevents haemolysis of RBC.

Procedure:

Mix thoroughly and place the tubes in boiling water bath for 30 minutes. Take the reading at 540 nm.

	Test	*Standard*	*Blank*
Distil Water	1.0 ml	-	-
PFF	-	-	1.0 ml
Standard (0.025 mg/dl)	-	1.0 ml	-
Working Reagent	5.0 ml	5.0 ml	5.0 ml

Calculation

$$\frac{\text{OD of test}}{\text{OD of std}} \times \frac{\text{Conc. of std}}{\text{Vol. of blood}} \times 100 = \text{mg of urea per 100 ml}$$

$$\frac{\text{OD of test}}{\text{OD of std}} \times \frac{0.025}{0.025} \times 100 = \text{mg of urea per 100 ml}$$

$$= \frac{T}{S} \times 100 \text{ mgs of urea per } 100 \text{ ml (mg/dl)}$$

$$\text{Millimols/liter} = \frac{\text{Mg percent} \times 10}{\text{Molecular weight (60)}}$$

BUN: Mol wt of urea is 60
2 atoms of N = 14 × 2 = 28

$$\text{Mg BUN} = \text{mg urea} \times \frac{28}{60}$$

$$= \text{mg urea} \div 2.14$$

Normal Range

BUN is 5 to 25 mg per 100 ml
Blood urea is 15 to 45 mg per 100 ml

Urea Clearance: Number of ml of blood that is cleared of urea that is excreted by the kidney in 1 min.

Types

a. Standard urea clearance–: The urine excretion is less than 2 ml/min
b. Maximal urea clearance–: The urine excretion is more than 2 ml/min

a. Standard urea clearance:

$$\frac{U\sqrt{v}}{P} = \frac{1.85U\sqrt{v}}{P}$$

N = 54 ml/min

b. Maximal urea clearance

$$= 1.33 \frac{Uv}{P}$$

N = 75 ml/min

Clinical Interpretation

Blood urea is commonly between 15 and 45 mg per 100 ml. Blood urea is lower in pregnancy than in normal non-pregnant women.

Increases in blood urea may occur in a number of diseases in addition to those in which the kidneys are primarily involved (Fig. 3.4).

1. **Pre-renal:**
   - Severe and protracted vomiting—pyloric and intestinal obstruction
   - Ulcerative colitis
   - Diabetic coma
   - Crisis of Addison's disease
   - Hematemesis
   - Severe burns
   - Toxic fever
   - Cardiac failure.

2. **Renal diseases:**
   - Acute glomerulonephritis
   - Malignant hypertension
   - Chronic pyelonephritis
   - Mercurial poisoning
   - Hydronephrosis
   - Congential cystic kidneys
   - Renal tuberculosis
   - Renal failure.
3. **Post-renal diseases:**
   - Enlargement of prostate
   - Stones in the urinary tract
   - Stricture of urethra
   - Tumors of the bladder affecting the ureters.

Decreased blood urea values have been reported in

- severe liver disease
- protein malnutrition, and
- pregnancy

Other Methods

a. Urease method.
   NH_3, Potassium mercuric iodide.

b. Phenol-hypochlorite method using the Berthelot reaction.

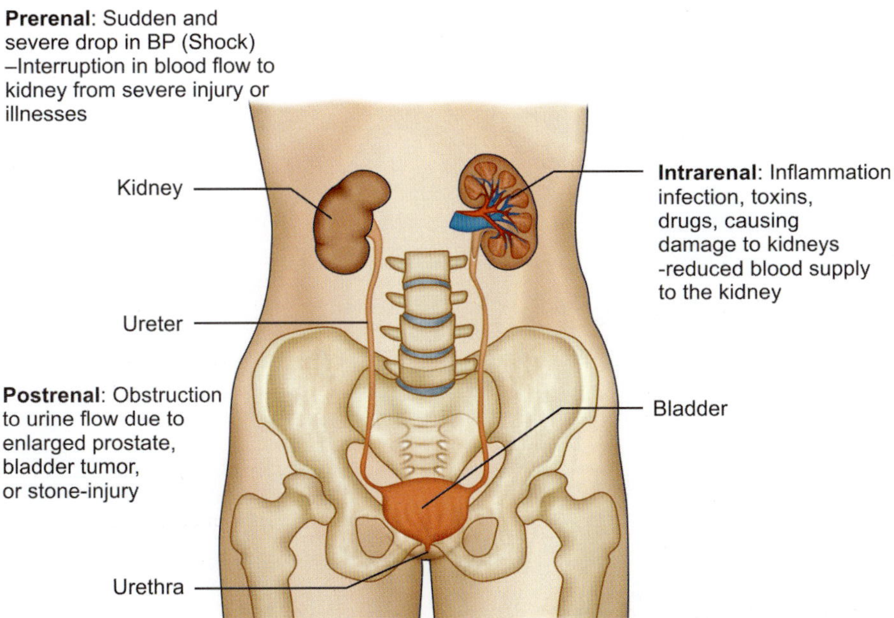

Fig. 3.4: Common causes of increased blood urea levels (Azotemia)

RELEVANT QUESTIONS

1. What is the normal blood urea level?
2. What is the normal amount of urea excreted in urine daily?
3. In which conditions urea level will be increased in blood?
4. Where is urea synthesized?
5. In which conditions urea level will be decreased in blood?'
6. What is urea clearance?
7. What are the other clearance tests?
8. Which is the most preferred clearance test and why?
9. What are the methods by which urea can be estimated in blood?
10. What is DAM?
11. What is the principle of DAM method?
12. What is the importance of "blank" and "Standard" in quantitative analysis?
13. What is "standard" and "maximal" urea clearance and their significance?
14. What is the difference between BUN and blood urea? What is the conversion factor?

ESTIMATION OF URINARY CREATININE

Creatinine is a nitrogenous waste product formed from the metabolism of creatine (as creatine phosphate) in skeletal muscle. The contractile protein actomyosin hydrolyzes ATP to ADP + P_i during muscle contraction. The ATP is rapidly restored by the Lohmann reaction during contraction. The depleted store of phosphocreatine is regenerated from ATP when the muscle is in resting state.

LOHMANN REACTION

$$\textbf{ADP + Phosphocreatine} \xrightarrow{\textbf{Creatine kinase}} \textbf{ATP + Creatine}$$

Creatinine diffuses freely throughout the body water. It is filtered from the extracellular fluid by the kidney and excreted in the urine. The excretion of creatinine is mainly renal and in absence of disease, is relatively constant.

Any disease or condition that causes a fall in the glomerular filtration rate (GFR) will increase plasma creatinine levels. Because creatinine is so readily excreted, blood levels rise more slowly than do urea levels in renal disease with slight increases occurring when there is moderate renal damage.

Normal Serum Creatinine Level

Males: 0.9: 1.3 mg/dl,
Females: 0.6: 1.1 mg/dl

Expt No 53: Estimation of Urinary Creatinine

Method: Alkaline picrate method or Jaffés method.

Aim: To estimate urinary creatinine in the given sample.

Reagents

1. Picric acid (0.04 m solution).
2. Sodium hydroxide (0.75 N solution)
3. Stock solution of creatinine containing 1 mg of creatinine per ml.
4. Working creatinine standard for use:
5. This contains 0.01 mg of creatinine per ml.

Principle: Creatinine in alkaline medium reacts with picric acid to form a yellowish orange tautomer of creatinine picrate, the intensity of which is measured at 520 nm. (About 2% of the total creatine is converted daily to creatinine so that the amount of creatinine produced is related to the total muscle mass.)

Procedure: Dilute 5 ml of urine to 500 ml in a volumetric flask (1:100 dilutions)

	Test	Standard	Blank
Diluted urine	-	-	3.0 ml
Creatinine std	-	3.0 ml	-
Distil water	3.0 ml	-	-
0.04 m picric	1.0 ml	1.0 ml	1.0 ml
0.75 N sodium hydroxide	1.0 ml	1.0 ml	1.0 ml

Allow to stand for fifteen minutes and then read in the colorimeter during the next half hour with a blue-green filter or at 520 nm

Calculations: Since the standard contains 0.03 mg creatinine and 3 ml of diluted urine corresponds to 0.03 ml of the original urine.
Grams Creatinine per litre of urine

$$= \frac{\text{OD of T} - \text{OD of B}}{\text{OD of S} - \text{OD of B}} \times 0.03 \times \frac{1000}{0.03} \times \frac{1}{1000}$$

$$= \frac{\text{OD of T} - \text{OD of B}}{\text{OD of S} - \text{OD of B}} \text{ gm/litre}$$

Note: In serum about 20% of the color is due to non-creatinine substances, whereas in urine it is about 5% due to noncreatinine substances. Nonspecific chromogens that react with picric acid are proteins, ketone bodies, pyruvate, glucose and ascorbate. Jaffe's reaction is also sensitive to variables like pH, temperature, etc.

Normal Range: The normal daily excretion

Males: 14 – 26 mg/kg/day

Females: 11 – 20 mg/kg/day.

(Declines with age to 10 mg/kg/day [88.4 μmol/kg/day] at age 90 yrs. Decline starts in fifth decade.)

Urinary creatinine 0.7–1.2 gm/day.

Interpretation

Urinary creatinine increases in:

- Exercise
- Acromegaly, gigantism
- Diabetes mellitus
- Infections
- Hypothyroidism
- Meat meals.

Urinary creatinine decreases in:

- Hyperthyroidism
- Anemia
- Paralysis, Muscular dystrophy
- Diseases with decreased muscle mass (e.g. Neurogenic Atrophy, Polymyositis, etc.),
- Inflammatory disease affecting muscle

- Advanced renal disease
- Leukemia
- Vegetarian diets.

Creatinine is not a sensitive indicator of early renal disease. Plasma creatinine is less affected than urea by dietary intake. For each 50% reduction in GFR, serum creatinine doubles. In chronic renal disease, the plasma level may be more sensitive to changes in glomerular function than creatinine clearance, which may be factitiously higher than the true value.

Creatinine is decreased unilaterally in urine from a kidney affected by renal stenosis.
Note: Determination of urine creatinine concentration is of little or no help in evaluation of renal function unless it is done as part of creatinine clearance test. Because the excretion of creatinine in one person is relatively constant (assuming a constant diet), 24-hr urine creatinine levels are used as an approximate check on the completeness of a 24-hr urine collection in serial collections.

Determination of Creatine in Urine

This is determined by estimating the creatinine present before and after heating with acid to convert the creatine into creatinine. From the difference the amount of creatinine can be calculated.

Calculation

1 gm of creatinine is formed from 1.16 gm of creatine. Subtract the preformed creatinine from the total creatinine after heating with acid and multiply by 1.16.

There is very little creatine in urine of adults, particularly in males. It may be found in the urine in pregnancy and also in childhood. Under certain abnormal dietary conditions it may occur in the urine of adults when a low carbohydrate diet is taken and in fasting when it is presumably derived from the breakdown of muscular tissues. It is also reported in increased amounts in the urine of athletes during training.

Creatinine Coefficient

The sum of the mgs of creatinine and creatine nitrogen excreted per kilogram of body weight is termed creatinine co-efficient.

Men: 7–10 mg/kg

Women: 5–8 mg/kg

Factors for converting creatinine and creatine into their nitrogen content are 0.372 and 0.321 respectively.

The glomerular filtration rate (GFR) is the rate at which an ultra filtrate of plasma is produced by glomeruli per unit of time. It is the best estimate of the number of functioning nephrons or functional renal mass.

Creatinine clearance, the most useful clinical estimate of GFR, is defined as the volume of plasma that is cleared of creatinine by the kidney per unit of time.

Creatinine clearance

$$\frac{\text{mg of creatinine excreted per minute}}{\text{mg of creatinine per ml of blood}} = \frac{UV}{P}$$

Where

U = mg of creatinine per 100 ml of urine

P = mg of creatinine per 100 ml of blood

V = ml of urine excreted per minute

Normal creatinine clearance:

Males: 95 – 140 ml/min. (Ave: 120 ml/min.)

Females: 85 – 125 ml/min. (Ave: 110 ml/min.)

Chronic kidney disease :< 60 ml/min/1.73 m^2

Kidney failure: <15 ml/min/1.73 m^2

Unlike precise GFR measurements involving constant infusions of inulin, creatinine is already at a steady state concentration in the blood, so measuring creatinine clearance is much less cumbersome. As the true GFR falls to 40 mL/min (as measured by inulin clearance), the absolute amount of creatinine secreted can rise by more than 50%, accounting for as much as 35% of urinary creatinine.

An alternative method for estimating GFR in patients with moderate to severe renal dysfunction is to take the average of the creatinine and urea clearances. The clearance of creatinine overestimates the GFR due to creatinine secretion, but the urea clearance underestimates GFR, as about 40% to 50% of the filtered urea is reabsorbed. Because the magnitude of the two errors tends to be similar, the average of both clearances is more accurate.

A more complete estimation of renal function can be made when interpreting the blood (plasma) concentration of creatinine along with that of urea. The ratio of blood urea nitrogen to creatinine can indicate other problems besides those intrinsic to the kidney; for example, a urea level raised out of proportion to the creatinine may indicate a pre-renal problem such as volume depletion.

The principle behind this ratio is the fact that both urea (BUN) and creatinine are freely filtered by the glomerulus, however urea is reabsorbed by the tubules, therefore can be regulated (increased or decreased) whereas creatinine reabsorption remains the same (minimal reabsorption). An elevated BUN: Cr due to a low or low-normal creatinine and a BUN within the reference range is unlikely to be of clinical significance.

Normal Serum Values

Test	*SI Units*	*US Units*
BUN	1.1 – 8.2 mmol urea/L	7 – 20 mg/dl
Urea	2.5 – 10.7 mmol/L	15 – 45 mg/dl
Creatinine	62 – 106 µmol/L	0.7 – 1.2 mg/dl

Interpretation of Serum Ratios of BUN, Urea and Creatinine

BUN:Cr	*Urea:Cr*	*Location*	*Mechanism*
>20:1	>100:1 40 – 100:1	Prerenal (before the kidney)	BUN reabsorption is increased. BUN is disproportionately elevated relative to creatinine in serum.
10 – 20:1		Normal or postrenal (after the kidney)	Normal range. Can also be due to postrenal disease. BUN reabsorption is within normal limits.
<10:1	<40:1	Intrarenal (within kidney)	Renal damage causes reduced reabsorption of BUN, therefore lowering the BUN: Cr ratio.

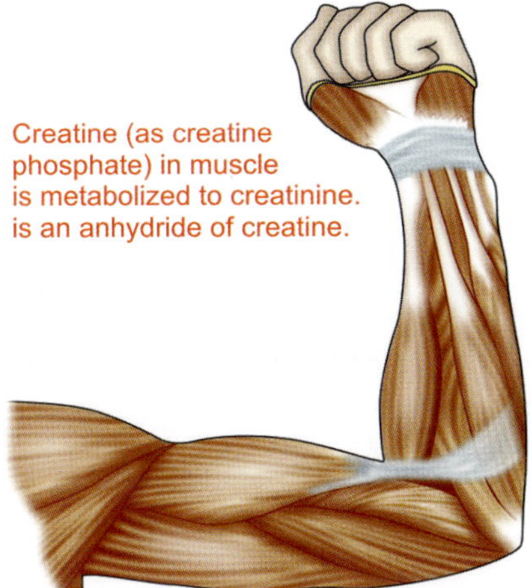

Conditions causing increased blood creatinine:
- High Blood Pressure
- Diabetes
- Hyperthyroidism
- Dehydration
- Kidney infection
- Muscle Conditions like muscular dystrophy, myasthenia gravis
- Urinary tract obstruction
- Acromegaly and Gigantism

Conditions causing increased urine creatinine:
- Strenuous exercise
- Muscle injury-crush injuries

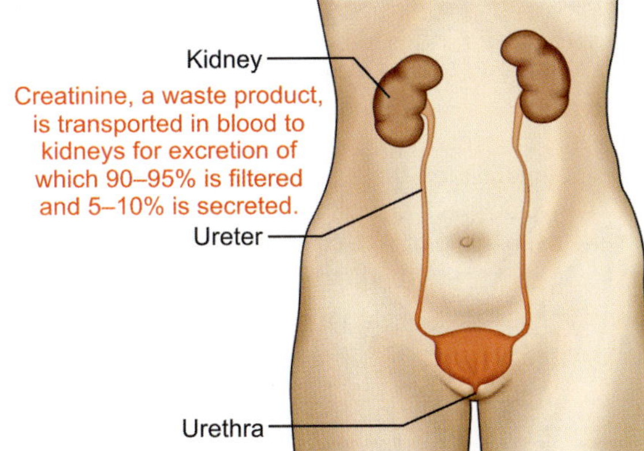

Conditions causing decreased blood creatinine:
- Debilitation due to increased age or decreased muscle mass
- Pregnancy

Conditions causing decreased urine creatinine:
- Kidney damage due to severe infection
- Shock
- Cancer
- Low blood flow to the kidneys
- Obstruction in the urinary tract
- Vegetarian diets

Fig. 3.5: Synthesis and excretion of creatinine and clinical conditions resulting in increased/decreased creatinine in blood and urine

RELEVANT QUESTIONS

1. What is creatinine?
2. What is the relationship between creatine and creatinine?
3. How is creatinine synthesized?
4. Where is creatinine synthesized?
5. What is the normal urinary creatinine level?
6. Name the conditions with increased urinary creatinine levels.
7. When will there be decreased urinary creatinine levels?
8. What is the normal serum creatinine level?
9. Name the conditions in which the serum creatinine level will be increased.
10. What is creatinine clearance? What is its normal value?
11. What are the other clearance tests?
12. Which is the most common clearance test performed in the clinical laboratory and why?
13. What is creatinine coefficient?
14. What is the normal urea: creatinine ratio in serum?
15. What is Lohmann reaction?

TOTAL SERUM PROTEIN ESTIMATION

Proteins are the most abundant compounds in serum (the rest of blood when you remove all the cells after clotting). Amino acids are the building blocks of all proteins. In turn proteins are the building blocks of all cells and body tissues. They are the basic components of enzymes, many hormones, antibodies and clotting agents. Proteins act as transport substances for hormones, vitamins, minerals, lipids and other compounds. In addition, proteins help to balance the osmotic pressure of the blood and tissue. Osmotic pressure is part of what keeps water inside a particular compartment of your body. Proteins play a major role in maintaining the delicate acid-alkaline balance of your blood. Finally, serum proteins serve as a reserve source of energy for your tissues and muscle when you are not ingesting an adequate amount.

The major measured serum proteins are divided into two groups, **albumin** and **globulins.** There are four major types of globulins, each with specific properties and actions. A typical blood panel will provide four different measurements—the total protein, albumin, globulins, and the albumin-globulin ratio.

Expt. No. 54: Estimation of Total Serum Proteins

Method: The Biuret method.
Aim: To estimate total serum protein in the given sample.

REAGENTS

1. Stock biuret reagent
2. Biuret solution for use
3. Tartrate iodide solution
4. Bovine or human albumin standard, 2 mg/ml.

Principle: Substances which contain two $CONH_2$ groups joined together directly or through a single carbon or nitrogen atom and those which contain two or more peptide links, give a purple colored compound with alkaline copper solution.

	Blank	*Standard*	*Test*
SERUM	-	-	0.1 ml
ALBUMIN	-	3.0 ml	-
DISTIL WATER	3.0 ml	-	2.9 ml
WORKING BIURET	3.0 ml	3.0 ml	3.0 ml

Procedure:

Mix and incubate all the tubes at room temperature for 10 minutes. Read absorbance against reagent blank at 540 nm or green filter.

Note: Hemolyzed blood increases the color intensity. 0.1 gm/dl increase for every 100 mg/dl of Hemoglobin.

Calculation:

$$\frac{T-B}{S-B}\times\frac{\text{Eff. Con. of std}}{\text{Eff. Vol of sample}}\times\frac{100}{1000}=\text{gm/dl}$$

$$=\frac{T}{S}\times\frac{6}{0.1}\times\frac{100}{1000}$$

$$=\frac{T}{S}\times 6=\text{gm/dl}$$

Other Methods for Estimating Proteins

1. Nitrogen estimation by Kjeldahl technique
2. Lowry's method for estimating tyrosine estimation

Interpretation

Normal Values:

Total serum proteins - 6–8 gm/dl

Albumin - 3.7–5.3 gm/dl

Globulin - 1.8–3.6 gm/dl

Fibrinogen - 200–400 mg/dl

The albumin-globulin ratio varies from about 2.5:1 to about 1.2:1 (1.5:1).

Total protein may be elevated due to:

- Chronic infection (including tuberculosis)
- Adrenal cortical hypofunction
- Collagen vascular disease (Rheumatoid arthritis, systemic lupus, scleroderma)
- Hypersensitivity states
- Sarcoidosis
- Dehydration (due to hemo concentration as in diabetic ketoacidosis, chronic diarrhea, etc.)
- Respiratory distress
- Hemolysis
- Cryoglobulinemia
- Alcoholism
- Leukemia

Total protein may be decreased due to:

- Malnutrition and malabsorption (insufficient intake and/or digestion of proteins)
- Liver disease (insufficient production of proteins)
- Diarrhea (loss of protein through the GI tract)
- Severe burns (loss of protein through the skin)
- Hormone imbalances that favor breakdown of tissue
- Loss through the urine in severe kidney disease (proteinuria)

- Low albumin (see "albumin")
- Low globulins (see "globulins")
- Pregnancy (dilution of protein due to extra fluid held in the vascular system)

ALBUMIN

Albumin is synthesized by the liver using dietary protein. Its presence in the plasma creates an osmotic force that maintains fluid volume within the vascular space. A very strong predictor of health; low albumin is a sign of poor health and a predictor of a bad outcome (Fig. 3.6).

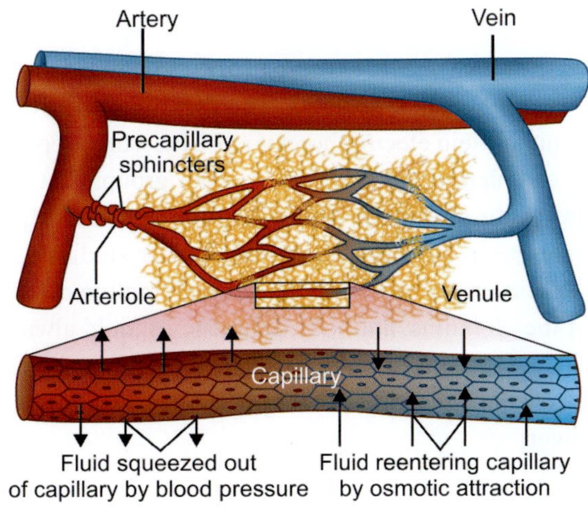

Fig. 3.6: Albumin maintains the osmotic pressure

Serum albumin levels may be elevated in:

- Dehydration - actual
- Congestive heart failure
- Poor protein utilization
- Glucocorticoid excess (can result from taking medications with cortisone effect, the adrenal gland overproducing cortisol, or a tumor that produces extra cortisol like compounds)
- Congenital

Serum albumin levels may be decreased in:

- Dehydration
- Hypothyroidism
- Chronic debilitating diseases (e.g. Rheumatoid arthritis)
- Malnutrition - Protein deficiency
- Dilution by excess H_2O (drinking too much water, which is termed "polydipsia," or excess administration of IV fluids)
- Kidney losses (Nephrotic syndrome)

- Protein losing-enteropathy (protein is lost from the gastrointestinal tract during diarrhea)
- Skin losses (burns, exfoliative dermatitis)
- In severe hemorrhage
- Liver dysfunction (the body is not synthesizing enough albumin and indicates very poor liver function)
- Insufficient anabolic hormones such as Growth Hormone, DHEA, testosterone, etc.

Congenital disorders: Analbuminemia, Disalbuminemia.

GLOBULINS

Globulins are proteins that include gamma globulins (antibodies) and a variety of enzymes and carrier/transport proteins. The specific profile of the globulins is determined by serum protein electrophoresis (SPEP), which separates the proteins according to size and charge. There are four major groups that can be identified: **gamma globulins, beta globulins, alpha-2 globulins, and alpha-1 globulins.** Once the abnormal group has been identified, further studies can determine the specific protein excess or deficit. Since the gamma fraction usually makes up the largest portion of the globulins, antibody deficiency should always come to mind when the globulin level is low. Antibodies are produced by mature B lymphocytes called plasma cells, while most of the other proteins in the alpha and beta fractions are made in the liver.

Optimal Range: 2.3 – 2.8 g/dL

Optimal Range (Alpha Globulin): 0.2 – 0.3 g/L

Optimal Range (Beta Globulin): 0.7 – 1.0 g/L

The serum globulin level may be elevated in:

- Chronic infections (parasites, some cases of viral and bacterial infection)
- Liver disease (biliary cirrhosis, obstructive jaundice)
- Carcinoid syndrome
- Rheumatoid arthritis
- Ulcerative colitis
- Multiple myelomas, leukemias, Waldenstrom's macroglobulinemia
- Autoimmunity (Systemic lupus, collagen diseases)
- Kidney dysfunction (Nephrosis)

The serum globulin level may be decreased in:

- Nephrosis (A condition in which the kidney does not filter the protein from the blood and it leaks into the urine)

- Alpha-1 Antitrypsin Deficiency (Emphysema)
- Acute hemolytic anemia
- Liver dysfunction
- Hypogammaglobulinemia/Agammaglobulinemia

ALBUMIN/GLOBULIN (A/G) RATIO

The liver can function adequately on 20% of liver tissue, thus early diagnosis by lab methods is difficult. A reversed A/G ratio may be a helpful indicator. With severe liver cell damage, the prolonged prothrombin time will not change with ingestion of Vitamin K. The proper albumin to globulin ratio is 2:1.

Optimal Range: 1.7 – 2.2

The A/G ratio may be elevated in:

- Hypothyroidism
- High protein/high carbohydrate diet with poor nitrogen retention
- Hypogammaglobulinemia (low globulin)
- Glucocorticoid excess (can be from taking medications with cortisone effect, the adrenal gland overproducing cortisol, or a tumor that produces extra cortisol like compounds, low globulin).

The A/G ratio may be decreased in:

- Liver dysfunction
- Multiple myeloma
- Cirrhosis
- Autoimmune disease
- Nephrotic syndrome

Albumin: Globulin ratio
Low A/G ratio due to overproduction of gamma-globulin monoclonal/polyclonal gammopathy, multiple myeloma or autoimmune diseases, etc. or due to low albumin (low production as in cirrhosis or excessive loss as in nephrotic syndrome or protein losing enteropathy, etc.).

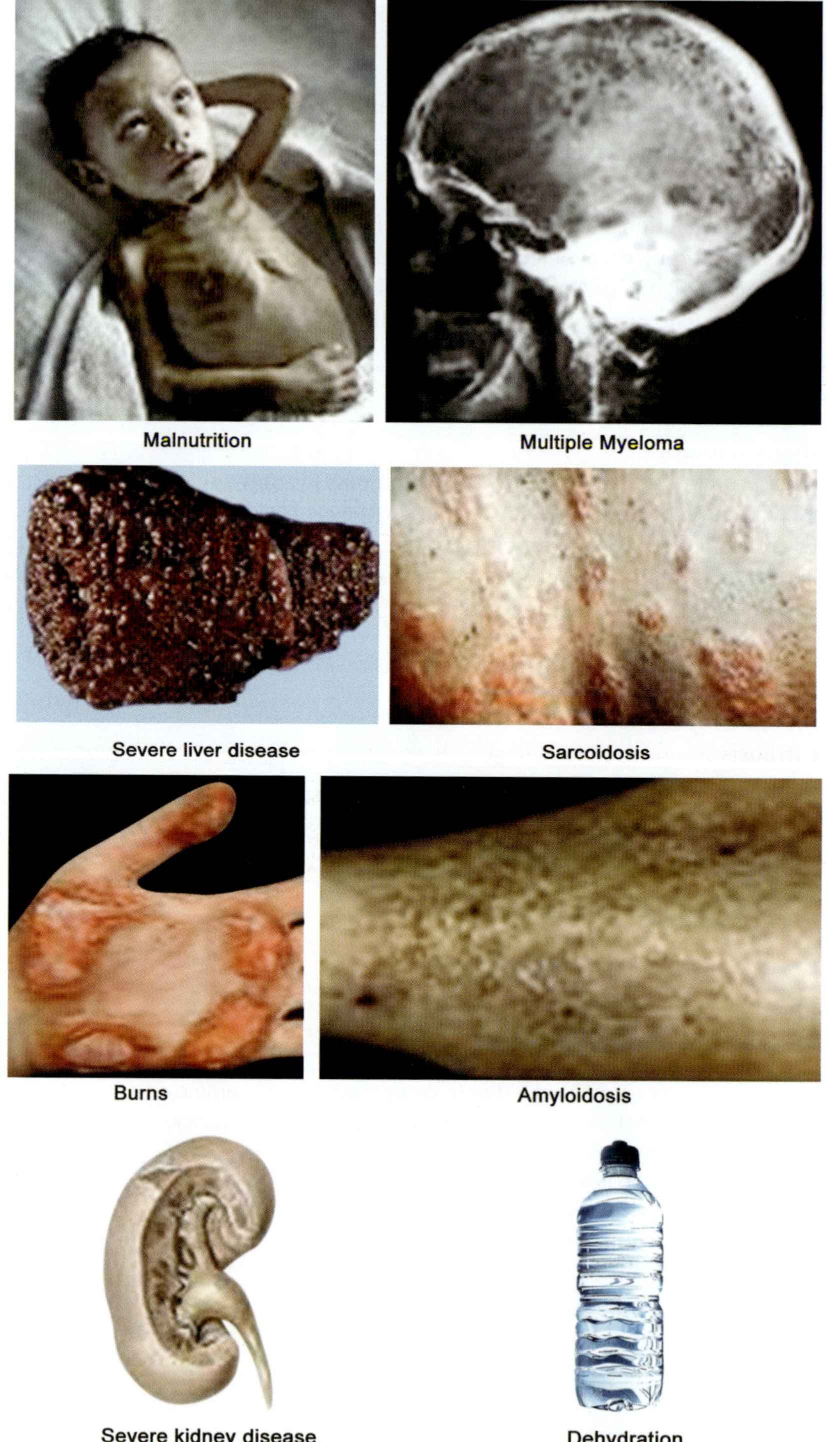

Fig. 3.7: Common causes of hypo- and hyperproteinemia

General Methods for Quantification of Proteins

Method	*Principle*	*Uses*
Kjeldahl's method	Measurement of protein content based on the nitrogen content. Protein content = nitrogen content x 100/16	Used as a reference method.
Biuret method	Proteins containing peptide bonds form a characteristic purple color when treated with dilute copper sulfate in alkaline solutions. The absorbance of the color is measured at 540 nm.	Used to determine concentration of total protein in the plasma.
UV absorption method.	Method based on absorption of UV light (280 nm) by aromatic amino acids (tyrosine and tryptophan) present in proteins.	Determining protein in a solution or elute from a chromatographic column.
Turbidometric method.	Method based on the precipitation of proteins using acids such as TCA or sulfosalicylic acid. Suspension of precipitated proteins blocking the light along a given path of light and the amount of the blocked light is proportional to the concentration of practicles.	Used to quantify proteins in cerebrospinal fluid and urine samples.
Lowry method Folin Ciocalteu assay	Method depends on the presence of tyrosine and trytophan content. The protein reacts with alkaline copper causing reduction of copper ions by tyrosine and trytophan residues. The cuprous ions, in turn, reduce the phosphomolybdate and phosphotungstate salts in the reagent to form an intensely blue complex.	Estimation of proteins especially during isolation of proteins.
Dye binding method (Bradford assay)	Method based on binding of proteins to Coomassie Brillant blue resulting shift in the absorbance of the dye from 465 nm to intense band at 595 nm.	More sensitive than Lowry assay.

RELEVANT QUESTIONS

1. What is the normal value of total proteins?
2. What are the normal values of albumin and globulin?
3. What is A:G ratio? What is its significance? What is the normal value?
4. What are plasma proteins? Name them.
5. What are the functions of plasma proteins?
6. How do you separate plasma proteins in the laboratory?
7. In which conditions plasma proteins level will be a) increased b) decreased?
8. What is reversal of AG ratio and what is its significance?
9. What are the causes of hypoproteinemias and hyperproteinemias?
10. What is plasma?
11. What is serum?
12. What is the difference between plasma and serum?
13. By which method protein is estimated?
14. What is the principle of biuret method?
15. What is Biuret and why is it named so?
16. What are the components of biuret reagent?
17. Name the methods by which plasma protein can be estimated in the laboratory?
18. What is the importance of "blank" and "standard" in quantitative analysis?

CEREBROSPINAL FLUID ANALYSIS

INTRODUCTION

The cerebrospinal fluid (CSF) is a metabolically active and dynamic substance which is an ultrafiltrate of plasma, secreted by the choroid plexus of the ventricles of brain. CSF then passes through the foramina of the fourth ventricle into the subarachnoid cisterns present at the base of the brain. It travels over the surface of the cerebral hemispheres before getting absorbed into the blood by the cerebral veins and dural sinuses. The volume of CSF in adults is 150 ml which is present in the cavity surrounding the brain (in the skull) and the spinal cord (in the spinal column).

CSF Performs the following Functions

1. Acts as a fluid buffer by protecting the brain and spinal cord from injury.
2. It provides nutrients to the brain and spinal cord.

Cerebrospinal fluid (CSF) analysis is a set of laboratory tests that examine a sample of the fluid surrounding the brain and spinal cord. This fluid is an **ultrafiltrate of plasma**. It is clear and colorless. It contains glucose, electrolytes, amino acids, and other small molecules found in plasma, but has very little protein and few cells. CSF protects the central nervous system from injury, cushions it from the surrounding bone structure, provides it with nutrients, and removes waste products by returning them to the blood. CSF is withdrawn from the subarachnoid space through a needle by a procedure called a **lumbar puncture or spinal tap. CSF analysis** includes tests in **clinical chemistry, hematology, immunology, and microbiology.**

Usually Three or Four Tubes are Collected

- The first tube is used for chemical and/or serological analysis.
- The last two tubes are used for hematology and microbiology tests.

This reduces the chances of a falsely elevated white cell count caused by a traumatic tap (bleeding into the subarachnoid space at the puncture site), and contamination of the bacterial culture by skin germs or flora.

Purpose of CSF Analysis

The purpose of a CSF analysis is to **diagnose medical disorders** that affect the central nervous system. Some of these conditions are:

- Meningitis and encephalitis, which may be viral, bacterial, fungal, or parasitic infections
- Metastatic tumors and central nervous system tumors that shed cells into the CSF
- Syphilis, a sexually transmitted bacterial disease
- Bleeding in the brain and spinal cord
- Multiple sclerosis, a degenerative nerve disease that results in the loss of the myelin coating of the nerve fibers of the brain and spinal cord

- Guillain-Barré syndrome, a demyelinating disease involving peripheral sensory and motor nerves

Routine examination of CSF includes **visual observation** of color and clarity and tests for glucose, protein, lactate, lactate dehydrogenase, red blood cell count, white blood cell count with differential count, syphilis serology (testing for antibodies indicative of syphilis), Gram stain, and bacterial culture.

Further tests may need to be performed depending upon the results of initial tests and the presumptive diagnosis. For example, an abnormally high total protein seen in a patient suspected of having a demyelinating disease such as multiple sclerosis dictates CSF protein electrophoresis and measurement of immunoglobulin levels and myelin basic protein.

Color and clarity are important diagnostic characteristics of CSF. Straw, pink, yellow, or amber pigments (xanthochromia) are abnormal and indicate the presence of bilirubin, hemoglobin, red blood cells, or increased protein concentration. Turbidity (suspended particles) indicates an increased number of cells. Gross examination is an important aid to differentiating a subarachnoid hemorrhage from a traumatic tap.

The latter is often associated with sequential clearing of CSF as it is collected; streaks of blood in an otherwise clear fluid; or a sample that clots.

Gross Examination

Gross appearance	*Cause*	*Significance*
Crystal clear		Normal
Smoky	RBC	Hemorrhage, Traumatic tap
Cloudy, Turbid	WBC, Micro-organisms, Protein	Meningitis, Disorders affecting blood-brain barrier, production of IgG within CSF
Bloody	RBC	Hemorrhage
Xanthochromic	Hemoglobin Bilirubin, Carotene	Old hemorrhage, increased serum bilirubin, Increased dietary carotene intake
Greenish tinge	Myelo-Peroxidase	Purulent fluid

Glucose: CSF glucose is normally approximately two-thirds of the fasting plasma glucose.

Low CSF Glucose Levels

- Bacterial Meningitis
- Fungal Meningitis
- Malignancy

Protein: Total protein levels in CSF are normally very low, and albumin makes up approximately two thirds of the total.

High CSF Total Protein Levels

- Bacterial meningitis
- Fungal meningitis

- Multiple sclerosis
- Tumors
- Subarachnoid hemorrhage
- Traumatic tap.

LACTATE

Increased CSF Lactate

- Bacterial and cryptococcal (fungal) meningitis
- In lactic acidosis (in disorders of gluconeogenesis, PDH complex, Krebs cycle and mitochondrial, etc.)

CSF lactate is normal in viral meningitis.

LACTATE DEHYDROGENASE

LDH – enzyme in CSF is elevated in

- Bacterial meningitis
- Fungal meningitis
- Malignancy
- Subarachnoid hemorrhage.

WHITE BLOOD CELL (WBC) COUNTS

The number of white blood cells in CSF is very low, usually necessitating a manual WBC count.

Increase in WBC:

- Infection (viral, bacterial, fungal, and parasitic)
- Allergy
- Leukemia
- Multiple sclerosis
- Hemorrhage
- Traumatic tap
- Encephalitis
- Guillain-Barré syndrome

The WBC differential count helps to distinguish many of the above causes. For example, viral infection is usually associated with an increase in lymphocytes, while bacterial and fungal infections are associated with an increase in polymorphonuclear leukocytes (neutrophils). Increase in differential eosinophils count is associated with allergy and ventricular shunts. Further, macrophages with ingested bacteria (indicating meningitis), RBCs (indicating hemorrhage), or lipids (indicating possible cerebral infarction); blasts (immature cells) that indicate leukemia; and malignant cells characteristic of the tissue of origin also help in diagnosis. About 50% of metastatic cancers that infiltrate the central nervous system and about 10% of central nervous system tumors will shed cells into the CSF.

RED BLOOD CELL (RBC) COUNTS

RBC is not normally found in CSF.

Appearance of RBC in CSF

- Subarachnoid hemorrhage
- Stroke
- Traumatic tap.

Since white cells may enter the CSF in response to local infection, inflammation, or bleeding, the RBC count is used to correct the WBC count so that it reflects conditions other than hemorrhage or a traumatic tap.

For every additional 700 RBCs seen due to presumed traumatic tap, one WBC is expected. (Normal WBC: RBC ratio = 1:700)

The predicted CSF WBC count can be calculated using the formula

$$\text{Predicted WBC}_{CSF} = \text{RBC}_{CSF} \times \frac{\text{WBC}_{blood}}{\text{RBC}_{blood}}$$

This is accomplished by counting RBCs and WBCs in both blood and CSF. The ratio of RBCs in CSF to blood is multiplied by the blood WBC count. This value is subtracted from the CSF WBC count to eliminate WBCs derived from hemorrhage or traumatic tap.

GRAM STAIN

The Gram stain is performed on sediment of the CSF and is positive in at least 60% of cases of bacterial meningitis. Culture is performed for both aerobic and anaerobic bacteria. In addition, other stains (e.g. the acid-fast stain for *Mycobacterium tuberculosis,* fungal culture, and rapid identification tests [tests for bacterial and fungal antigens]) may be performed routinely.

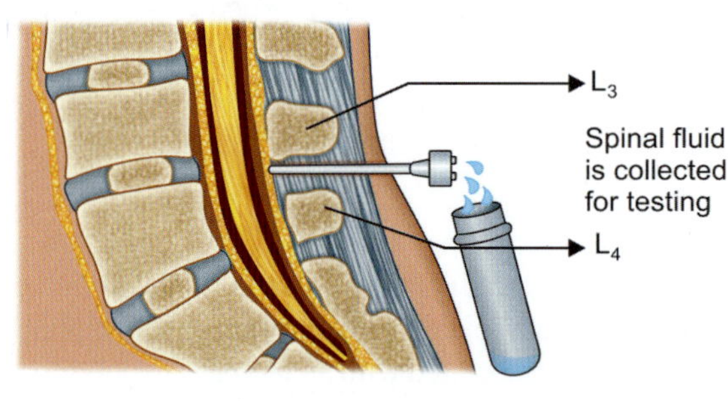

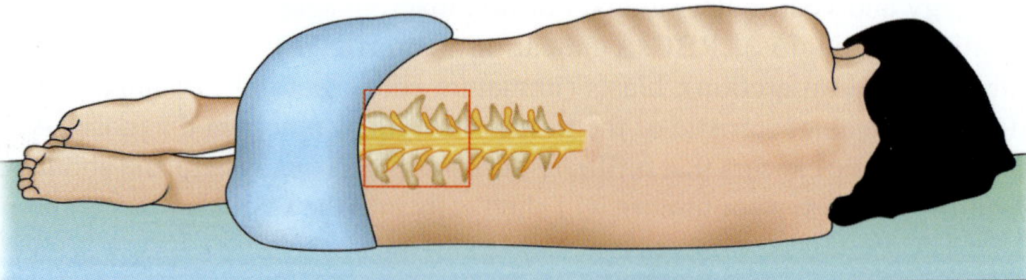

Cerebrospinal fluid analysis

SYPHILIS SEROLOGY: This involves testing for antibodies that indicate neurosyphilis. The fluorescent treponemal antibody-absorption (FTA-ABS) test is often used and is positive in persons with active and treated syphilis. The test is used in conjunction with the VDRL test for nontreponemal antibodies, which is positive in most persons with active syphilis, but negative in treated cases.

Normal Composition of CSF

1. Color : Colorless
2. Appearance : Clear
3. CSF opening pressure : 50 – 175 mm H_2O
4. pH : 7.3 – 7.4
5. No clot formation on standing
6. Specific gravity : 1.003 – 1.008
7. Total solids : 0.85 – 1.70 g/dl

   Protein : 15 to 45 mg/dl
(Albumin = 50 – 70%)
(Globulins = 30 – 50%)

   Glucose : 40–80 mg/dl

   Chlorides : (700 – 750 mg of NaCl/100 ml)
120 – 130 mEq/1

   Sodium : 144 – 154 mEq/1

   Potassium : 2.0 – 3.5 mEq/1

   Creatinine : 0.5 – 1.2 mg/dl

   Cholesterol : 0.2 – 0.6 mg/dl

   Urea : 6 – 16 g/dl

- Uric acid : 0.5 – 4.5 mg/dl
- LD : 1/10 of serum level
- Lactate: Less than 35 mg/dL.
- Leukocytes (white blood cells): 0–5/microL (adults and children); up to 30/microL (newborns).
- Differential: 60–80% lymphocytes; up to 30% monocytes and macrophages; other cells 2% or less. Monocytes and macrophages are somewhat higher in neonates.
- Gram stain: Negative.
- Culture: Sterile.
- Syphilis serology: Negative.
- Red blood cell count: Normally, there are no red blood cells in the CSF unless the needle passes through a blood vessel on route to the CSF.

Specimen Collection

Under aseptic conditions, the sterile lumbar puncture needle is inserted between the 3th and 4th lumbar vertebrae to a depth of 4 to 5 cm. This location is used because the spinal cord stops near L2, and a needle introduced below this level will miss the cord.

After the removal of stylet the fluid is collected through the needle into two/four test tubes.

Tube 1 – About 0.5 ml of CSF—for serological analysis.

Tube 2 – About 3 to 5 ml of CSF. The CSF is centrifuged and the supernatant used for following biochemical tests.

Tube 3 – About 1 ml of CSF—hematology tests.

Tube 4 – About 0.5-1 ml of CSF—microbiology tests.

Expt. No. 55. Cerebrospinal Fluid Analysis

(1) CSF glucose

(2) Determination of CSF total protein.

Method

CSF protein estimation by turbidimetry method.

Principle: Proteins are amphoteric in nature, i.e. they behave as acids in an alkaline medium and as bases in an acidic medium. In the presence of sulphosalicylic acid (alkaloid reagents), they act as bases, and react with the acid to form an insoluble salt of protein sulphosalicylate.

Reagents

1. Sulphosalicylic acid – 3 gm/dl sulphosalicylic acid
2. Normal saline – 0.85 gm/dl of sodium chloride
3. Stock protein standard – 6 gms/dl of bovine albumin stable for one year when refrigerated.
4. Working protein standard – 60 mg/dl – Prepare by mixing 0.1 ml of the stock standard and 9.9 ml of normal saline. This should be prepared fresh.

Procedure: Pipette in the tubes labeled as follows:

	Test	*Std*	*Blank*
Sulphosalycylic acid (3 gm/dl)	4.0 ml	4.0 ml	4.0 ml
CSF	1.0 ml	-	-
Working standard (60 mg/dl)	-	1.0 ml	-
Distill water	-	-	1.0 ml

Keep at room temperature for 5 minutes and take the reading at 450 mμ.

Calculations

$$\frac{\text{OD of T} - \text{OD of B}}{\text{OD of S} - \text{OD of B}} \times \frac{\text{Eff. Conc. of STD}}{\text{Eff. Vol of sample}} \times 100 = \text{mg/dl}$$

$$\frac{\text{T}}{\text{S}} \times 60 = \text{mg/dl}$$

QUALITATIVE TESTS FOR GLOBULIN

Two tests are in use for showing the presence of an increase in globulin in CSF:

1. The Nonne-Apelt test
2. Pandy's test

1. NONNE-APELT TEST:
The globulin is precipitated by half-saturating with ammonium sulfate.

Reagent

Saturated ammonium sulfate solution. Dissolve 85 gm of ammonium sulfate in 100 ml of hot distilled water, allow to cool overnight and filter. Keep in a well stoppered bottle.

Procedure: Add 1 ml of this saturated solution to 1 ml of fluid while shaking stand for three minutes and read the amount of opalescence on turbidity by inspection. Report findings as:
- No opalescence
- Slight opalescence
- Opalescence
- Marked opalescence, and
- Turbidity

Almost all normal fluids remain quite clear and never show more than the faintest suspicion of opalescence.

2. PANDY'S TEST:

Reagent

Pandy's solution: Prepare by dissolving 10 gms of phenol in 150 ml of distilled water. The solution should be clear and colorless.
Procedure: Add 2 drops of CSF to 2 ml of Pandy's solution. Most normal fluids show no opalescence at all.
The test is reported as:
- No opalescence
- Slight opalescence
- Opalescence
- Marked opalescence, and
- Turbidity.

The test has the advantage of requiring only 2 drops of fluid and is more sensitive than the Nonne-Apelt.

Expt. No 55 (3). Determination of CSF Chlorides:
Method: Titration
Reagents
1. Silver nitrate solution – 30 mEq/L
3. Potassium chromate –10% solution

Procedure: Pipette 1 ml of CSF into a conical flask and add 2 ml of distil water. Add two drops of potassium chromate and titrate with silver nitrate to faint brick red color (Normal CSF fluids require 4 to 4.35 ml).

Calculation

mEq/L of chlorides = ml of silver nitrate solution required X 30

Recap of Changes Observed in CSF in Various Clinical Conditions

Clinical Condition	*Appearance*	*Cells/Cumm (µL)*	*Glucose*	*Chlorides*	*Proteins*
Bacterial infection	Cloudy	>500 Neutrophils	Low values (0–40 mg/dl)	Marked decrease 600–700 mg/dl	High values (45–500 mg/dl) Increase in globulin
Viral infection	Clear	(10–200) mostly lymphocytes	Slightly low or normal	Moderate decrease	High values (45–300 mg/dl)
Fungal infection (very rare) Cryptococcus neoformans	Clear	(0–5) lymphocytes	Low values (0–40 mg/dl)	Normal or slight decrease	Normal
Acute purulent meningitis	Cloudy to purulent clot	Very high count (500–20,000) per cumm mostly neutrophils	Very low values (0–40 mg/dl)	Low values (600–700 mg/dl)	Very high (45–1000 mg/dl). Increase in globulins
Tuberculous meningitis	Cloudy, fibrin web	High count (10–500) mostly lymphocytes	Very low values (0–40 mg/dl)	Very low values (500–600 mg/dl)	High values (45–500 mg/dl). Increase in globulin
Acute syphilitic meningitis	Clear or Turbid	High count (20–2000) mostly lymphocytes	Low values (0–40 mg/dl)	Normal of slightly decreased	Normal, globulin: normal
Brain tumor	Clear or Turbid	High count (25-2000) mostly lymphocytes	Low values (0–40 mg/dl)	Normal	Increased, globulin: normal
Brain tumor	Clear	0–5	Increased	Normal	Increased, globulin: increased
Cerebral hemorrhage	Xantho-Chromic	0–5	Variable	Normal	Increased, globulin: normal
Encephalitis Lethargica	Clear	10-100 all lymphocytes	Slightly increased 80–120 mg/dl	Normal	normal or increased

RELEVANT QUESTIONS

1. What is CSF?
2. How is CSF collected?
3. What is the normal volume of CSF in adults?
4. What is the composition of CSF?
5. What is the method for estimating CSF glucose?
6. What is the normal CSF glucose concentration?
7. What is the method for estimating CSF proteins and what is the normal level?
8. What is the method for estimating CSF chloride and what is its normal range?
9. Name the conditions presenting with increased CSF proteins.
10. In which conditions will there be decreased CSF glucose level?
11. Name the qualitative tests for CSF proteins? Which is the better test and why?
12. What is the principle for CSF chloride determination test?

SERUM BILIRUBIN

INTRODUCTION

Bilirubin is synthesized in the reticuloendothelial cells from **hemoglobin present in RBC** and **transported in conjugation with albumin** (the normal indirect reacting bilirubin). On arrival at the sinusoidal surface of a hepatocyte, bilirubin separates from albumin and binds to a carrier protein. Within **hepatocyte, bilirubin is conjugated** and thereafter excreted in bile as bilirubin diglucuronide (normal direct reacting bilirubin when reacted with diazotized sulfanilic acid or Ehrlich's reagent). The azobilirubin thus formed is determined photometrically. There are **2-distinct forms of reaction**:

1. **Fast or direct**: Given by aqueous soluble conjugated bilirubin. Develops within 1 minute.
2. **Slow or indirect:** Given by insoluble albumin bound bilirubin. Reacts very slowly on longstanding.

Both forms of bilirubin are soluble in methyl alcohol.

ESTIMATION OF SERUM BILIRUBIN

Method: Malloy and Evelyn

Principle: Bilirubin reacts with diazotized sulphanilic acid to form purple colored azobilirubin. The water soluble bilirubin glucuronides (conjugated bilirubin) react fast (within one minute) with the diazo reagent (direct reaction). The free bilirubin (unconjugated bilirubin) which is present in serum complexed with albumin reacts very slowly and requires an accelerator or solubilizer as methanol (indirect reaction).

REAGENTS

1. Diazo reagent: Make fresh before use by mixing 10 ml of solution A and 0.3 ml of solution B.

Solution A: 1 gm sulfanilic acid and 15 ml of concentrated HCl per liter in water. This solution can be kept indefinitely at room temperature.

Solution B: 0.5 gm sodium nitrite per 100 ml in water. This solution should be kept in the Refrigerator.

2. Diazo blank: 15 ml of conc. HCl per liter in water.
3. Methyl red standard: Dissolve 0.29 gm pure methyl red in glacial acetic acid and make to 100 ml with the acid. To 1 ml, add 5 ml acetic acid and add 14.4 gm of sodium acetate (as trihydrate), dissolve and make to a liter with water. OD of this artificial standard corresponds to 10 mg/dl bilirubin.

Procedure

Mix and wait for 30 minutes and read the OD at 540 nm or green filter using distill water as blank.

	Total Bilirubin		*Direct Bilirubin*	
	Test	***Blank***	***Test***	***Blank***
Serum	0.2 ml	0.2 ml	0.2 ml	0.2 ml
Standard	-	-	-	-
Distill water	1.8 ml	1.8 ml	4.3 ml	4.3 ml
Diazo reagent	0.5 ml	-	0.5 ml	-
Diazo blank	-	0.5 ml	-	0.5 ml
Methanol	2.5 ml	2.5 ml	-	-

Calculation

1. Direct bilirubin (mg/ dl) = $\frac{\text{OD of D test} - \text{OD of D blank}}{\text{OD of std}} \times 10$

2. Total bilirubin (mg/dl) = $\frac{\text{OD of total test} - \text{OD of total blank}}{\text{OD of std}} \times 10$

Precautions

1. Hemolyzed sample should not be used because hemoglobin interferes with the diazo reaction and it also absorbs at 540 nm.
2. Serum for bilirubin estimation must be kept away from bright light since bilirubin is destroyed by ultraviolet light. Therefore, all tubes should be kept in dark.

Interpretation

Normal total bilirubin: 0.2 – 1.2 mg/dl
Indirect bilirubin: 0.2 – 0.7 mg/dl
Direct bilirubin: 0.1 – 0.4 mg/dl
Hyperbilirubinemia of more than 3 mg/dl results in clinical jaundice.

Clinical Interpretation

Jaundice is a condition presenting with yellowish discoloration of the skin, sclera, and mucous membranes due to increase in the concentration of bilirubin circulating in blood. Other causes which mimic jaundice are substances like carotene and certain drugs.

Conjugated bilirubin gets easily bound to elastin and other tissues high in protein content.

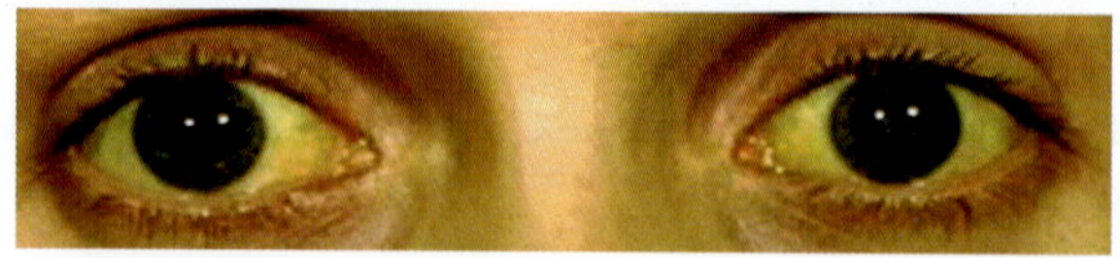

Overt jaundice is observed in patients when the bilirubin concentration rises above 2.0 to 3.0 mg/dl. Jaundice is classified as:

- Pre-hepatic or Hemolytic
- Hepatic or Hepatocellular
- Post-hepatic or Obstructive

Determination	*Pre-hepatic*		*Hepatic*		*Post-hepatic*
	Hemolytic	*Congential*	*Infective, toxic*	*Intra-hepatic cholestatic*	*Extra-hepatic cholestatic*
			Serum		
Total bilirubin	++++	+	+++	+++	+++
Direct bilirubin	N	N	++	++	++
Indirect bilirubin	+++	+	++	++	+
			Urine		
Bile pigments	N/-	N	++	++	++
Bile salts	N	N	++	++	++
Urobilinogen	++++	+	+	+	N or –

Key: N: Normal (Fig. 3.8) +++: High
+: Increased ++++: Very high
++: Moderately increased –: Absent

Types of Jaundice

1. ***Hemolytic Jaundice or Prehepatic (Fig. 3.9):***
   In hemolytic (prehepatic) jaundice, more bilirubin is produced than the liver can metabolize, e.g. in severe hemolysis (breakdown of red cells). The excess bilirubin which builds up in the plasma is mostly of unconjugated type and is therefore not found in the urine. Thus also called acholuric jaundice.
2. ***Hepatocelluar Jaundice or Hepatic (Fig. 3.10):***
   In hepatocellular (hepatic) jaundice, there is a build up of bilirubin in the plasma because it is not transported, conjugated, or excreted by the liver cells because they are damaged, e.g., in viral hepatitis. The excess bilirubin is usually of both the unconjugated and conjugated types with bilirubin being found in the urine.
3. ***Obstructive Jaundice or Posthepatic (Fig. 3.11):***
   In obstructive (posthepatic) jaundice, bilirubin builds up in the plasma because its flow is obstructed in the small bile channels or in the main bile duct. This can be caused by gallstones or a tumor obstructing or closing the biliary tract. The excess bilirubin is mostly of the conjugated type and is therefore found in the urine. The term cholestasis is used to describe a failure of bile flow.

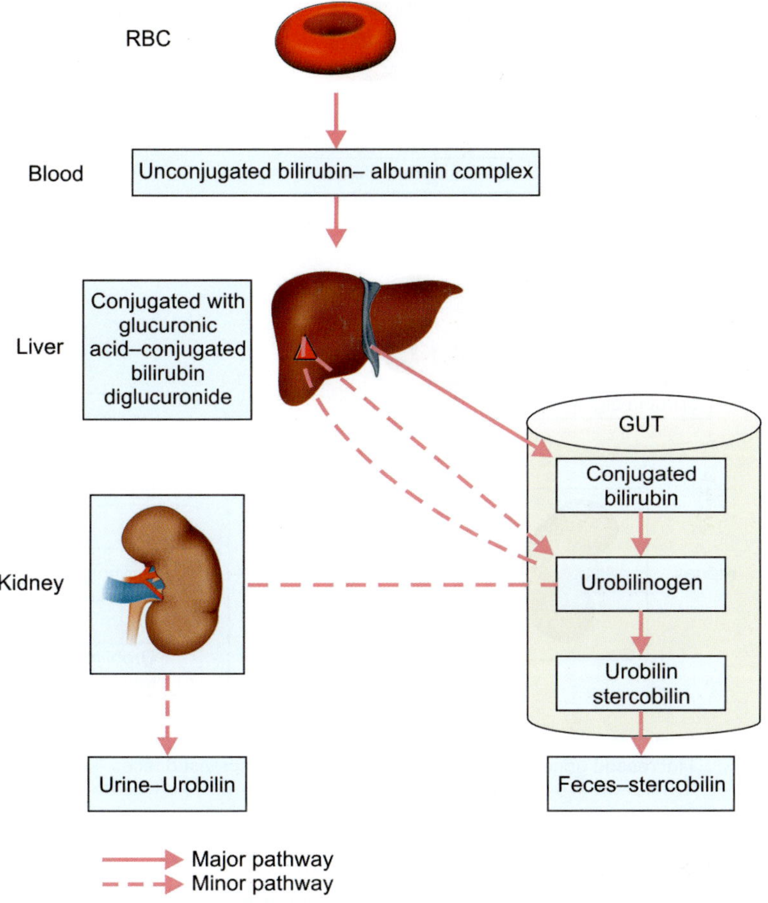

Fig. 3.8: Normal bilirubin metabolism

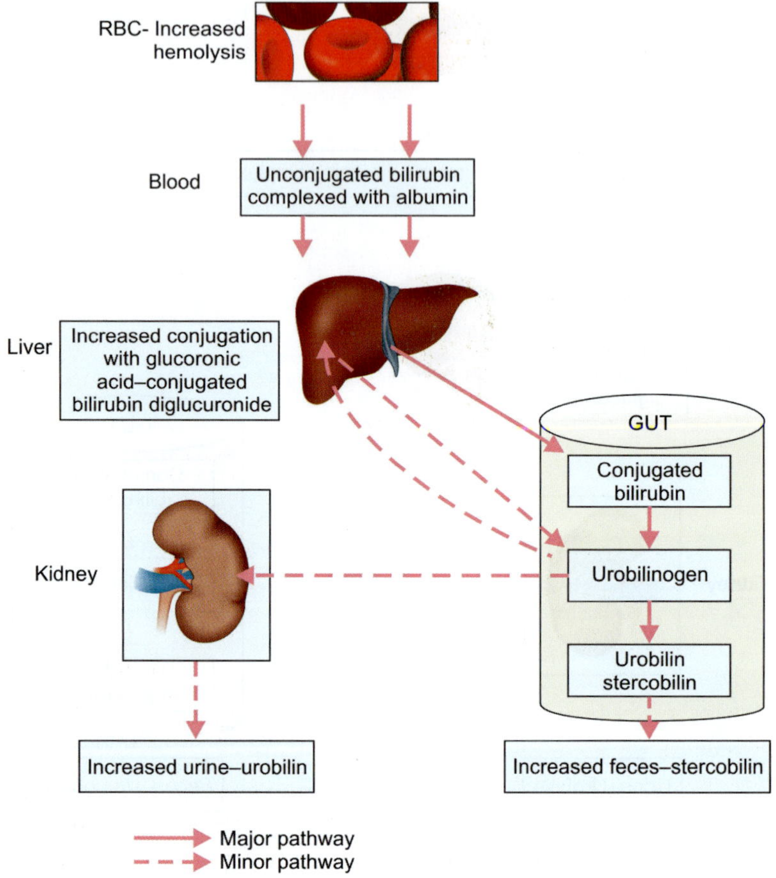

Fig. 3.9: Prehepatic jaundice

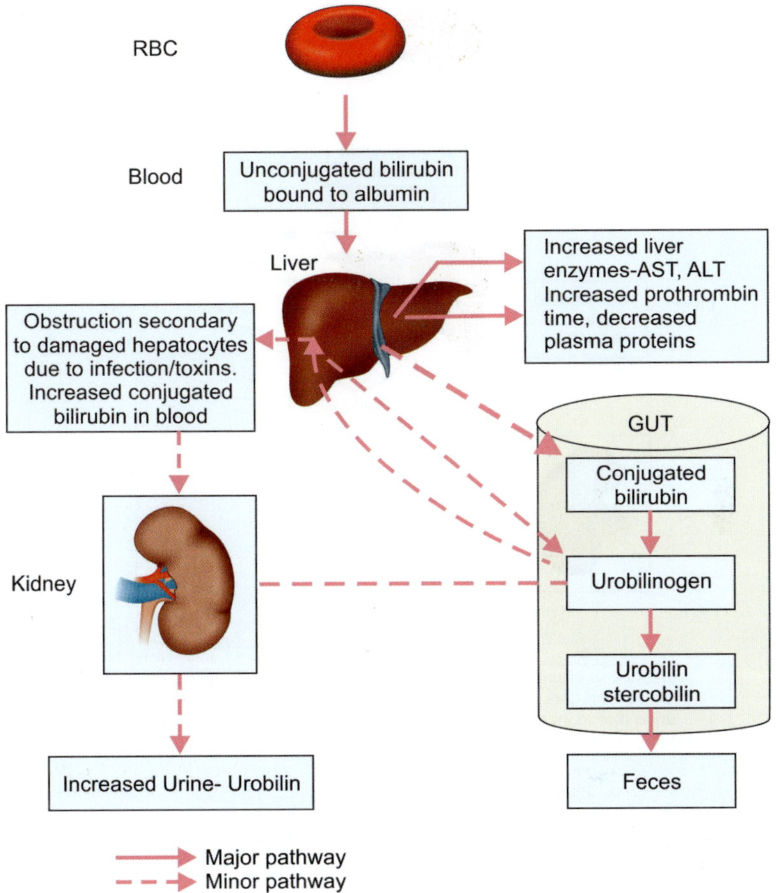

Fig. 3.10: Hepatic jaundice

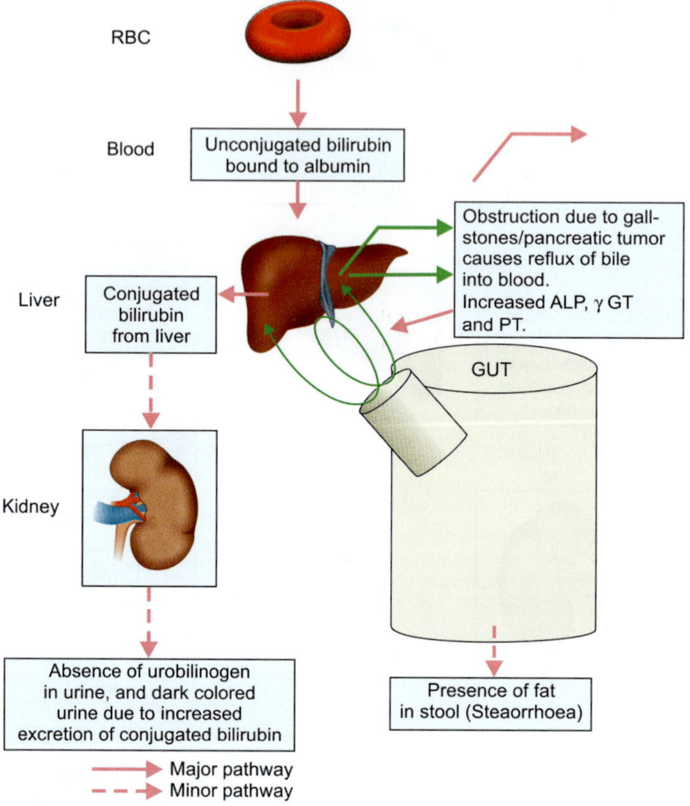

Fig. 3.11: Post hepatic jaundice

ESTIMATION OF SERUM URIC ACID

Aim: Determination of serum uric acid.

INTRODUCTION

The breakdown product of purines in man is uric acid. The catabolism of purine nucleotides is accomplished as given below. Adenosine is first deaminated to form inosine. This reaction is catalyzed by adenosine deaminase which is found in liver and other tissues. Inosine is converted to hypoxanthine by the action of nucleoside phosphorylase. Guanosine is converted to hypoxanthine by the action of nucleoside phosphorylase. Guanosine is converted to guanine which is then deaminated to form xanthine. This conversion is catalyzed by guanase which is found in liver, spleen, pancreas and kidneys. Hypoxanthine is oxidized to form xanthine by xanthine oxidase which also catalyzes the formation of uric acid from xanthine. In man, purine metabolism ends as uric acid which is excreted in urine. In mammals other than primates, however, uric acid is further metabolized to form allantoin by the enzyme uricase.

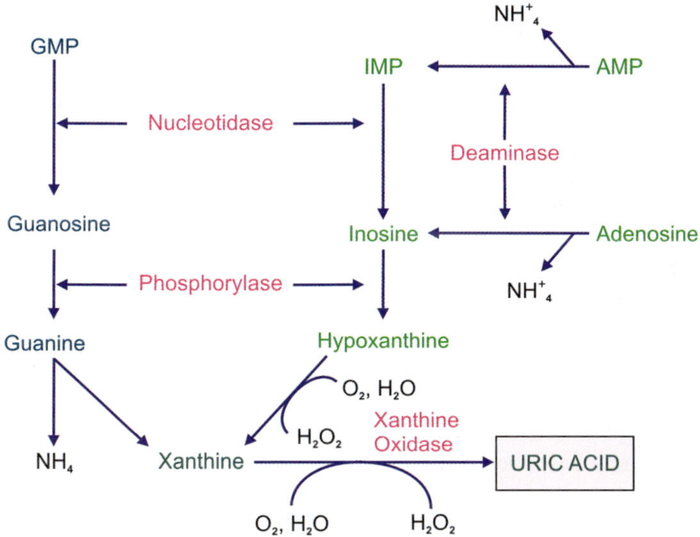

Method: Henry-Caraway's method.

Principle: Uric acid in the protein free filtrate reacts with phosphotungstic acid reagent in the presence of sodium carbonate (alkaline solution) to form a blue colored complex. The intensity of the color is measured at 660 nm (red filter).

Reagents

1. Deproteinizing reagent
   a. 10 g/dl, sodium tungstate: 50 ml
   b. 2/3 N, sulfuric acid: 50 ml

   c. Orthophosphoric acid: One drop
   d. Distilled water: 800 ml

   Mix well and store at room temperature in an amber colored bottle.

2. 10 gm/dl (w/v): Sodium carbonate

   Dissolve 10 gms of Na_2CO_3 in distilled water and make up to 100 ml with distilled water. This reagent is stable at room temperature when stored in a polythene reagent bottle.

3. Stock phosphotungstic acid reagent
   a. Sodium tungstate (Molybdate free): 50 gms
   b. Orthophosphoric acid: 40 ml
   c. Distilled water: 400 ml

   Mix and reflux gently for two hours. Cool and make final volume up to 500 ml. Store at 2–8°C in an amber colored container.

Note: Working phosphotungstic acid reagent is prepared fresh by diluting the stock solution 1:10.

4. Stock uric acid standard: 100 mg/dl

   Heat about 80 ml of distilled water in 250 ml beaker to 60°C. Add 60 mg, lithium carbonate and mix well. Add 100 mg uric acid and mix thoroughly. Add 2 ml formalin and then slowly with shaking 1 ml (1:2) acetic acid mix well and make final volume 100 ml by adding distilled water. Store in an amber colored bottle at 2 – 8°C.

Sample: Serum

Procedure:

1. Dilute the stock phosphotungstic acid, 1:10 by mixing 1.0 ml of the reagent and 9.0 ml of distilled water. Mix well
2. Dilute the stock uric acid standard (1:200) 0.1 ml of standard 100 mg/dl and 19.9 ml of distilled water. Mix well.
3. Pipette into a centrifuge tube labeled as follows:

	Test
Deproteinizing reagent, ml	5.4
Serum, ml	0.6

   Mix thoroughly and centrifuge at 3000 RPM for 10 minutes.

4. Pipette in the tubes labeled as follows:

	Test	*Standard*	*Blank*
Filtrate, ml	3.0	-	-
Diluted standard, ml	-	3.0	-
Distilled water, ml	-	-	3.0
Na_2CO_3 reagent, ml	1.0	1.0	1.0
Diluted phosphotungstic acid, ml	1.0	1.0	1.0

Mix, keep in the dark for exactly 10 minutes. Read OD of test and standard at 660 nm (red filter) against blank.

B = 0.00
S = 0.05
T = 0.04

Calculations

Serum uric acid, mg/dl

$$= \frac{O.D_T - O.D_B}{O.D_S - O.D_B} \times \frac{\text{Conc.of Std}}{\text{Vol.of Serum}} \times 100$$

$$= \frac{O.D_T - O.D_B}{O.D_S - O.D_B} \times \frac{0{,}015}{0.3} \times 100$$

$$= \frac{T}{S} \times 5 = \frac{0.04}{0.05} \times 5 = 4\text{mg/dl}$$

Procedure Limitation: This method is linear up to 10 mg/dl value of uric acid. For example, values above 10 mg/dl dilute the sample 1:2 in distilled water and repeat the assay.

Clinical Significance

Uric acid is end product of nucleoproteins metabolism. The total uric acid production in a normal subject has been estimated to be about 500 mg/day. Nearly 85% of the total uric acid produced is excreted as such in urine and feces. It is a low threshold excretory product. Uric acid is cleared from the plasma both by glomerular filtration and tubular secretion. The excretion of uric acid is increased by administration of adrenal cortex hormones ACTH and also by certain uricosuric drugs, e.g. salicylates. The serum uric acid level is often raised in gout. The determination has diagnostic value in differentiating gout from nongouty arthritis. Uric acid levels are also increased in renal failure, uremia and leukemia.

Normal Values:

Serum uric acid: 2 – 5 mg/dl for females
2 – 7 mg/dl for males.

Hyperuricemia may occur because of **decreased excretion** (underexcretors), **increased production** (overproducers), or a combination of these two mechanisms.

Underexcretion

- **Idiopathic**
- **Familial juvenile gouty nephropathy:** This is a rare autosomal dominant condition characterized by progressive renal insufficiency. These patients have a low fractional excretion of urate (typically 4%). Kidney biopsy findings indicate glomerulosclerosis and tubulointerstitial disease but no uric acid deposition.
- **Renal insufficiency:** Renal failure is one of the more common causes of hyperuricemia. In chronic renal failure, the uric acid level does not generally become elevated until the creatinine clearance falls below 20 mL/min, unless other contributing factors exist. This is due to a decrease in urate clearance as retained organic acids compete for secretion in the proximal tubule. In certain renal disorders, such as medullary cystic disease and chronic lead nephropathy, hyperuricemia is commonly observed even with minimal renal insufficiency.

- **Syndrome X:** This metabolic syndrome is characterized by hypertension, obesity, insulin resistance, dyslipidemia, and hyperuricemia. This is associated with a decreased fractional excretion of urate by the kidneys.
- **Drugs**: Causative drugs include diuretics, low-dose salicylate, cyclosporine, pyrazinamide, ethambutol, levodopa, nicotinic acid, and methoxyflurane.
- **Hypertension**
- **Acidosis:** Types that cause hyperuricemia include lactic acidosis, diabetic ketoacidosis, alcoholic ketoacidosis, and starvation ketoacidosis.
- **Preeclampsia and eclampsia:** The elevated uric acid associated with these conditions is a key clue to the diagnosis because uric acid levels are lower than normal in healthy pregnancies.
- **Hypothyroidism**
- **Hyperparathyroidism**
- **Sarcoidosis**
- **Lead intoxication (chronic):** History may reveal occupational exposure, e.g. lead smelting, battery and paint manufacture, or consumption of moonshine i.e. illegally distilled corn whiskey because some, but not all, moonshine was produced in lead-containing stills.
- **Trisomy 21**

Overproduction

- **Idiopathic**
- **HGPRT deficiency (Lesch-Nyhan syndrome):** This is an inherited X-linked disorder. HGPRT catalyzes the conversion of hypoxanthine to inosinic acid, in which PRPP serves as the phosphate donor. The deficiency of HGPRT results in accumulation of PRPP, which accelerates purine biosynthesis with a resultant increase in uric acid production. In addition to gout and uric acid nephrolithiasis, these patients develop a neurologic disorder that is characterized by choreoathetosis, spasticity, growth, mental function retardation, and occasionally, self-mutilation.
- **Partial deficiency of HGPRT (Kelley-Seegmiller syndrome):** This is also an X-linked disorder. Patients typically develop gouty arthritis in the second or third decade of life, have a high incidence of uric acid nephrolithiasis, and may have mild neurologic deficits.
- **Increased activity of PRPP synthetase:** This is a rare X-linked disorder in which patients make mutated PRPP synthetase enzymes with increased activity. These patients develop gout when aged 15 – 30 years and have a high incidence of uric acid renal stones.
- **Purine-rich diet:** A diet rich in meats, organ foods, alcohol, and legumes can result in an overproduction of uric acid.
- **Increased nucleic acid turnover:** This may be observed in persons with hemolytic anemia and hematologic malignancies such as lymphoma, myeloma, or leukemia.

- **Tumor lysis syndrome:** This may produce the most serious complications of hyperuricemia.
- **Glycogen storage diseases III, V, and VII.**

Combined Causes

- **Alcohol :** Ethanol increases the production of uric acid by causing increased turnover of adenine nucleotides. It also decreases uric acid excretion by the kidneys, which is partially due to the production of lactic acid.
- **Exercise:** Exercise may result in enhanced tissue breakdown and decreased renal excretion due to mild volume depletion.
- **Deficiency of aldolase B (fructose-1-phosphate aldolase)**: This is a fairly common inherited disorder, often resulting in gout.
- **Glucose-6-phosphatase deficiency (glycogenosis type I, von Gierke disease)**: This is an autosomal recessive disorder characterized by the development of symptomatic hypoglycemia and hepatomegaly within the first 12 months of life. Additional findings include short stature, delayed adolescence, enlarged kidneys, hepatic adenoma, hyperuricemia, hyperlipidemia, and increased serum lactate levels.

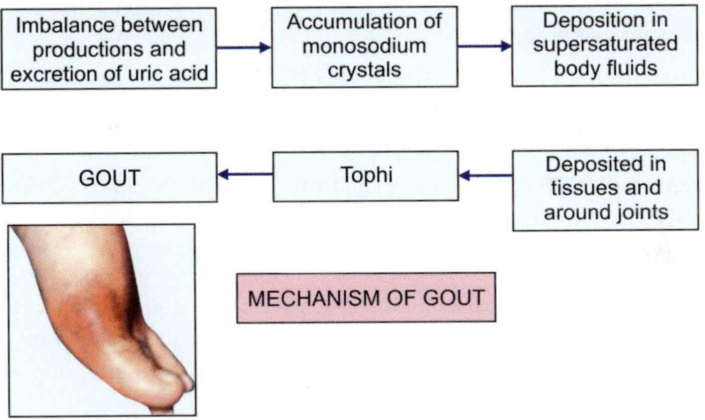

DEMONSTRATION EXPERIMENTS

GLUCOSE TOLERANCE TEST (GTT)

Glucose tolerance: A GTT measures the ability of the body to tolerate, metabolize or cope with a standard dose of glucose. The degree of tolerance to the glucose, as shown by a change in the blood level, is mainly dependent on the rate of glucose absorption and on the insulin response. As the glucose is absorbed, the level of glucose in the blood rises and the normal response is for insulin to be released from pancreas to lower the glucose level. Tolerance is reduced when insulin is insufficient or absent.

A glucose tolerance test **(GTT)** is usually **indicated** in:

- Fasting plasma glucose concentration is between 100 – 140 mg/dl
- Random plasma concentration is between 140 – 200 mg/dl
- Presence of glycosuria or
- A high index of clinical suspicion

If glycosuria is found, measurement of fasting glucose should be performed before the patient is subjected to a GTT. It allows the diagnosis of impaired fasting glycemia (IFG), impaired glucose tolerance (IGT) and gestational diabetes between 24 and 28 weeks of pregnancy.

Types of Glucose Tolerance Tests

1. Oral glucose tolerance test (OGTT)
2. Intravenous glucose tolerance test (IGTT)

Preparation of Patient

- To eat a balanced diet that contains at least 150 to 200 gms of carbohydrate per day for 3 days before the test. Fruits, breads, cereals, grains, rice, crackers, and starchy vegetables such as potatoes, beans, and corn are good sources of carbohydrate.
- Overnight fast of 8–14 hours and allowed to drink only water.
- Instructed not to eat, drink, smoke, or exercise strenuously for at least 8 hours before first blood sample is taken.
- The patient is instructed to stop taking certain medicines before the test. Several drugs may cause glucose intolerance, including the following:
  - Thiazide diuretics (e.g., hydrochlorothiazide)
  - Beta-blockers (e.g., propranolol)
  - Oral contraceptives
  - Corticosteroids (e.g., prednisone)
  - Some psychiatric medications
- Adults ingest 75 g glucose in 250–300 ml water over 5 minutes to 15 minutes. To reduce nausea a few drops of lemon juice may be added to the water.
- Children ingest 1.75 g/kg body weight in a similar volume of water by ratio (max 75 g as for adults).

Procedure

The most common glucose tolerance test is the oral glucose tolerance test (OGTT). Blood is taken for the determination of the fasting blood glucose and a specimen of urine is collected. Blood for estimation of blood glucose is taken at half hourly intervals, for 2½ hrs to 3 hrs after the glucose has been drunk. Urine specimens are collected at same 1/2 hr intervals or (two specimens) 1 hr and 2 hrs after taking the glucose, can usually be obtained.

The intravenous glucose tolerance test (IGTT) is not often used and is indicated in individuals with malabsorption or inability to tolerate oral glucose load. In this test, patients are given given 0.3 gm/Kg body weight in 30% solution intravenously and blood glucose is estimated every 3 minutes for 60 minutes. Blood insulin levels are measured before glucose and at 1 and 3 minutes. Insulin levels below a standard threshold may predict the development of type 1 diabetes in some patients.

Interpretation

Normal blood values for a 75 gram oral glucose tolerance test used to check for type 2 diabetes:

- Fasting: 60 – 100 mg/dL
- 1 hour: Less than 200 mg/dL
- 2 hours: Less than 140 mg/dL. Between 140–200 mg/dL is considered impaired glucose tolerance (sometimes called "prediabetes"). This group is at increased risk for developing diabetes. Greater than 200 mg/dL is a sign of diabetes mellitus.

Normal blood values for a 50 gram oral glucose tolerance test used to screen for gestational diabetes:

- 1 hour: Equal to or less than 140 mg/dL.

Normal blood values for a 100 gram oral glucose tolerance test used to screen for gestational diabetes:

- Fasting: Less than 95 mg/dL
- 1 hour: Less than 180 mg/dL
- 2 hours: Less than 155 mg/dL
- 3 hours: Less than 140 mg/dL

1999 WHO Diabetes Criteria - Interpretation of Oral Glucose Tolerance Test

Glucose levels	*Normal*		*Impaired fasting glycemia (IFG)*		*Impaired fasting tolerance (IFT)*		*Diabetes mellitus (DM)*	
Venous plasma	Fasting	2 hrs	Fasting	2 hrs	Fasting	2 hrs	Fasting	2 hrs
mg/dl	<110	<140	≥110 & <126	<140	<126	≥ 140	≥ 126	≥ 200
mmol/dl	<6.1	<7.8	≥6.1 & <7.0	<7.8	<7.0	≥7.8	≥7.0	≥11.1

TYPES OF GLUCOSE TOLERANCE

1. Normal glucose tolerance
2. Abnormal glucose tolerance
   - Decreased glucose tolerance
   - Increased glucose tolerance

Decreased glucose tolerance—in which the blood glucose level peaks acutely before declining slowly than usual to normal levels–as in:

- Diabetes mellitus
- Hemochromatosis (Iron overload disease)
- Cushing syndrome
- Pheochromocytoma
- Central nervous system lesions.

Increased glucose tolerance–in which the blood glucose level rise at slower than normal rate – as in:

- Malabsorption syndrome
- Insulinoma
- Addison's disease
- Hypopituitarism
- Hypothyroidism

Normal

1. Fasting plasma glucose: Less than 110 mg/dl
2. Peak plasma glucose: Less than 160 mg/dl
3. 2-hr plasma glucose: Less than 140 mg/dl
4. Urine glucose: Absent in all the specimens

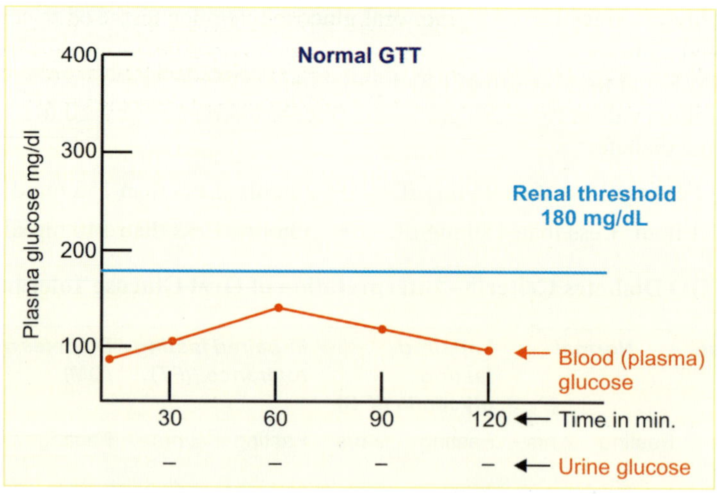

a. Normal response:

Initial zero hour fasting blood glucose value is within the normal range and the maximum blood glucose value reaches either ½ hour or 1 hour after taking the glucose. The blood glucose then returns rapidly to the normal fasting limits which often reaches in 1½ hours and almost always at 2 hours. By then, it should be below 120 mg%. Urine samples collected during test should be free of glucose.

Impaired glucose tolerance (IGT)

1. Fasting plasma glucose: Less than 126 mg/dl.
2. Peak plasma glucose: Less than 199 mg/dl.
3. 2-hr plasma glucose: Between 140-199 mg/dl.
4. Urine glucose: May be absent or present.

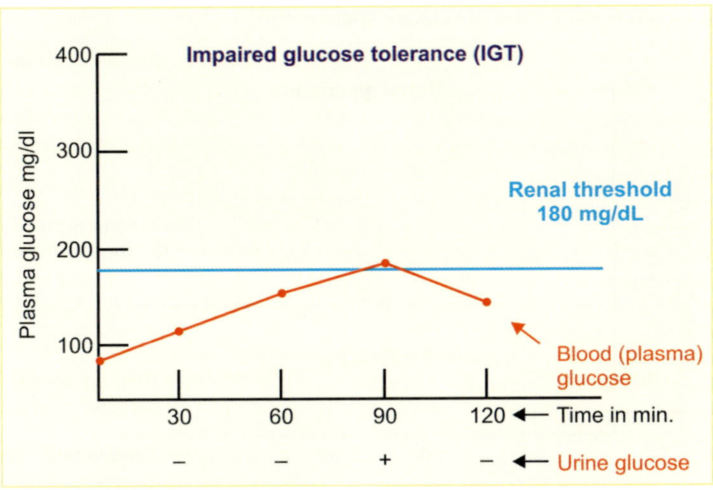

b. Abnormal curves

i. ***Impaired glucose tolerance:*** Glucose tolerance diminishes to a greater extent in diabetes mellitus. The most significant finding in the diagnosis of diabetes is the failure of the blood glucose level to fall below 120 mg even by 2 hours. The peak level is frequently above normal curve and fasting level may or may not be raised.

ii. ***Renal glycosuria:*** The curve is normal but one or more samples of urine contains glucose. When the blood levels are elevated, the glomerular filtrate may contain more glucose than can be reabsorbed, so the excess passes into the urine to produce glycosuria. Normally glycosuria occurs when venous blood conc exceeds 180 – 200 mg%. This is termed as renal threshold for glucose. Carbohydrate is taken to raise their blood glucose temporarily above their lowered renal threshold.

Carbohydrate metabolism is normal but because of some abnormality in the tubular reabsorption of glucose, an appreciable amount of glucose escapes in the urine. This condition is called renal glycosuria. This type of curve is harmless and patients are not likely to develop diabetes.

iii. ***Lag type of tolerance curve:*** Some otherwise normal individuals show an exaggerated rise in blood glucose, following an oral load but the level quickly falls and the 2 hrs concentration is within normal limits. Transient glycosuria usually occurs. This phenomenon probably results from an increased rate of glucose absorption from the gut, following rapid emptying of the stomach and occurs in some hyperthyroid patients. The increase in blood glucose level is due to delay in insulin mechanism coming into action.

Renal glycosuria

1. Fasting plasma glucose: Less than 110 mg/dl.
2. Peak plasma glucose: Less than 160 mg/dl.
3. 2-hr plasma glucose: Less than 140 mg/dl.
4. Urine glucose: Present in all the specimens.

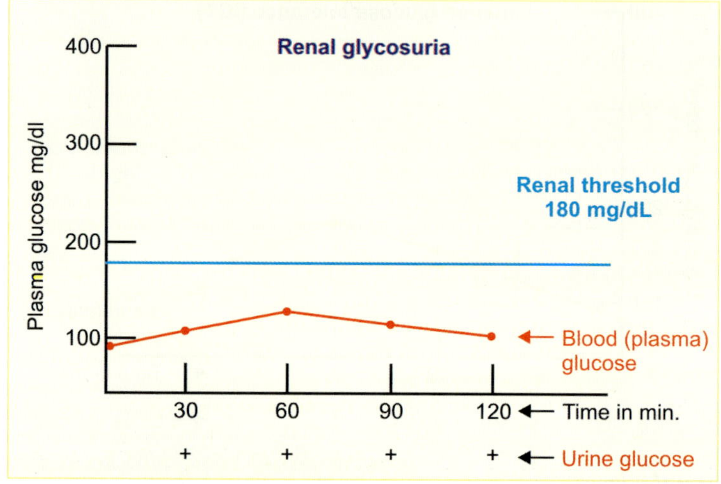

Postprandial Hypoglycemia

1. Fasting plasma glucose: Less than 110 mg/dl.
2. Peak plasma glucose: Less than 160 mg/dl.
3. 2-hr plasma glucose: Less than 70 mg/dl.
4. Urine glucose: May be present in one of the specimens

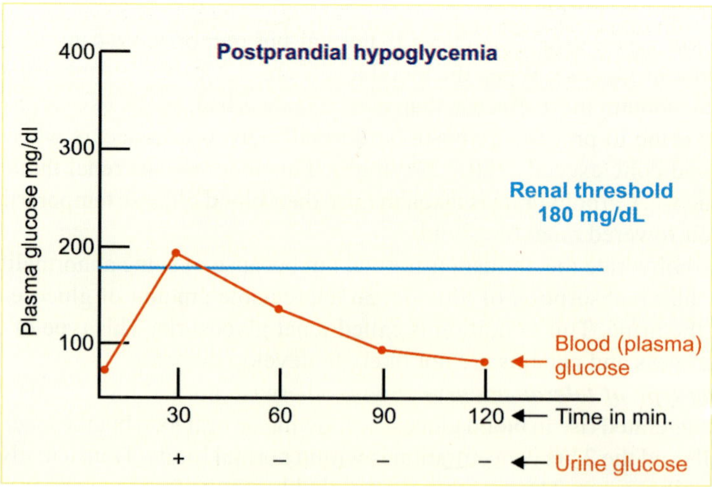

iv. ***The flat curve of enhanced glucose tolerance:*** In normal subjects given a glucose load, blood glucose levels rise to a peak at 30 or 60 min and then fall to near-fasting levels at 2 hrs but some individuals show very little rise. Flat curves are seen in patients with hypoactivity of other endocrine organs, e.g. in hypopituitarism and Addison's disease but also in malabsorption and in a proportion of normal subjects. Intravenous GTT may be performed in malabsorption patients.

Fasting hypoglycemia

1. Fasting plasma glucose: Less than 70 mg/dl.
2. Peak plasma glucose: Less than 160 mg/dl.
3. 2-hr plasma glucose: Less than 70 mg/dl.
4. Urine glucose: Absent in all the specimens.

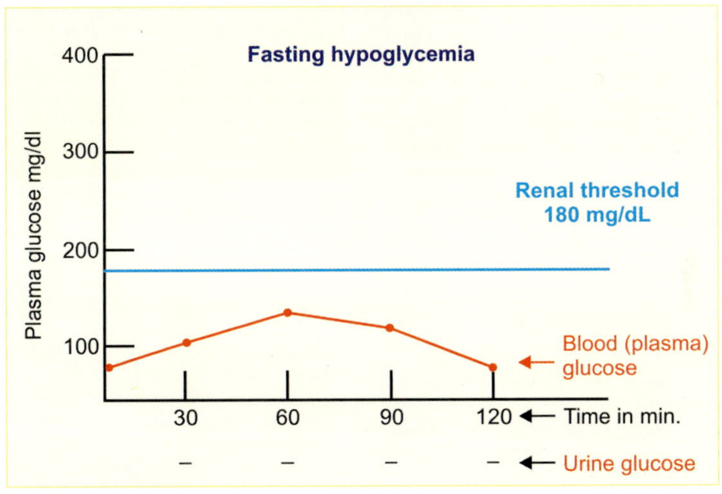

Diabetes mellitus

1. Fasting plasma glucose: More than 126 mg/dl.
2. Peak plasma glucose: More than 200 mg/dl.
3. 2-hr plasma glucose: More than 200 mg/dl.
4. Urine glucose: Present in some samples.

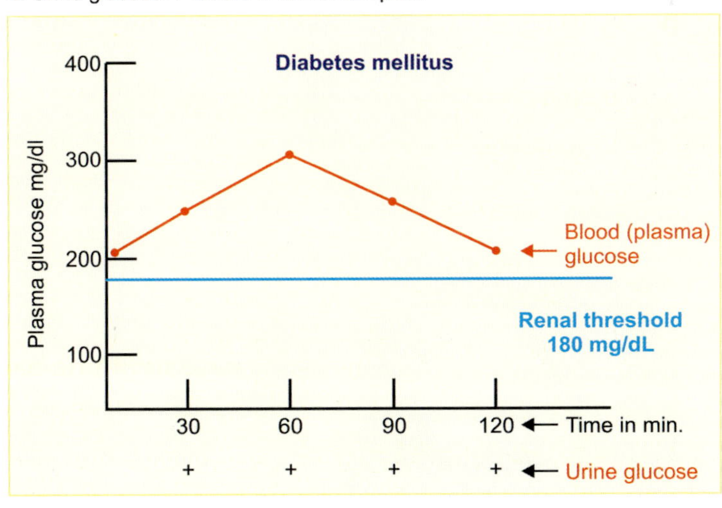

v. ***Diabetes mellitus curve:*** In diabetic patients, the fasting glucose level is high and further rises higher than a normal individual because the diabetic patient is unable to efficiently utilize the administered glucose load due to insufficient insulin supply. Glucose levels remain higher for longer period before returning to the initial level. According to WHO diagnostic criteria, a fasting glucose level of more than 126 mg/dl and a 2-hour sample of 200 mg/dl is diagnostic of diabetes mellitus.

Electrophoresis

Electrophoresis is a comprehensive term that refers to the migration of charged particles in a liquid medium under the influence of electric field. It is a versatile and powerful analytical technique used to separate and analyze a diverse range of ionized analytes.

ANALYTES SEPARATED BY ELECTROPHORESIS

- Proteins and peptides
- Nucleotides and nucleosides
- Organic acids
- Hemoglobin variants
- Lipoproteins
- Isoenzymes, etc.

Principle: An ampholyte or Zwitter ion which is neither positively nor negatively charged, takes positive charge in acidic media and negative charge in alkali media. Hence in electric field they migrate towards cathode (positive charged) and anode (negative charged) respectively.

The rate of migration depends upon–

1. Net electric charge
2. Size and shape of molecule
3. Electrical field strength
4. Properties of supporting media
5. Temperature

Electrophoretic mobility is defined as rate of migration (cm/s) for unit field strength (volts/cm)

Q $\rightarrow$ net charge
r $\rightarrow$ ionic radius
n $\rightarrow$ viscosity of buffer
$\mu \rightarrow$ electrophoretic mobility

$$\mu = \frac{Q}{6\pi r\eta}$$

Electrophoretic mobility is directly proportional to charge and inversely proportional to size and viscosity.

Procedure:

1. Instrument: A typical electrophoretic apparatus should have –
   1. Buffer chamber: two in number which has platinum electrodes, connected to power supply.
   2. An electrophoretic support media on which separation takes place which is in contact with buffer solution.

The entire apparatus is covered to minimize the evaporation of buffer.

SCHEMATIC DIAGRAM OF AN ELECTROPHORESIS APPARATUS

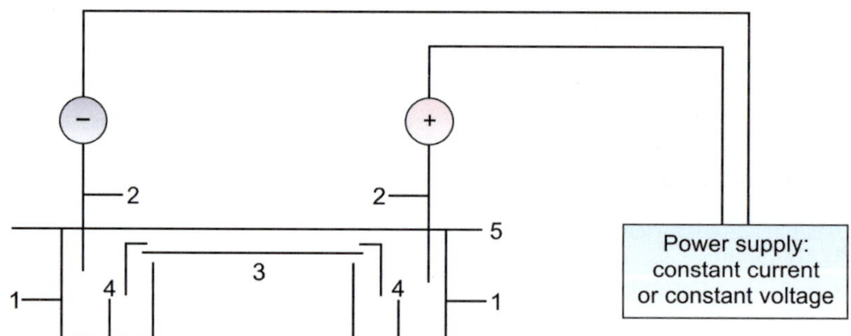

1. Buffer chamber 2. Connecting wire to the power source 3. Support media 4. Wicks dipping into the buffer (for completing the electrical circuit) 5. Lid

1. **Support media:**
   Gels are more commonly used–Agarose, cellulose acetate, polyacrylamide gel
   a. **Agarose:** Agar has agaropectin, a charged acidic polysaccharide with monomers of sulfated galactose and carboxylic side groups. Agarose is a purified, sulfate free fraction of Agar. It is free of ionizable groups and is therefore neutral. Pore size is large and independent of concentration when compared to molecular size of proteins. Hence separation is based only on charge to mass ratio of the proteins.
   Advantages: Lower affinity to proteins, clarity after drying, can be stored indefinitely and excellent in densitometric determination.
   b. **Cellulose acetate:** Cellulose is treated with acetic anhydride; the OH groups are acetylated to from cellulose acetate. Cellulose acetate has small pore size and hence is used regularly for hemoglobin electrophoresis.
   Advantages: Better resolving power than paper and better speed.
   c. **Polyacrylamide gel:** Polymer of acrylamide. Several layers of gels of different pore sizes are used. This gel has a high loading capacity and is routeinly used to separate DNA and individual proteins like isoenzymes. Gels are uncharged and are commercially available with variety of pore sizes.

2. **Power supply**: The function is to supply power during the electrophoresis process. It allows regulation of conditions like operation at constant voltage, current or power which are all adjustable as per the requirements.
3. **Buffers:** Buffers can affect the charge on the ampholyte by change in pH and ionic strength.

Multifunctional components:

1. Carries the applied current
2. Establish the pH
3. Determine the electrical charge of the solute.

Commonly used: Barbitone buffer (pH = 8.6)
Tris Boric Acid buffer

4. **Fixative:** 90% acetone
5. **Stains:** Used to visualize and locate the separated analytes.
   Serum proteins → Amidoschwarz, Bromophenol blue, Lissamine green.
   Isoenzymes → Nitrotetrazolium
   Lipoproteins → Sudan black; oil Q red.
   DNA → Ethidium bromide
   CSF proteins → $AgNO_3$
6. **Destaining agent:** 2% glacial acetic acid
   Automated systems: Commercially available automated systems are available with prepacked gels.

Types of Electrophoresis

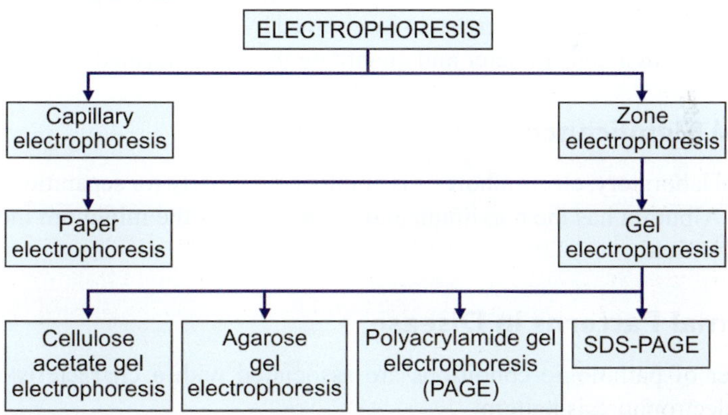

Advanced Techniques of Electrophoresis Now in Routine Use

1. Immunoelectrophoresis
2. Isoelectric focusing
3. Blotting techniques

General Procedure for Serum Protein Electrophoresis

1. **Preparation of slides:** Dissolve 200 mg of pure agarose in 20 ml of barbitone buffer and boil in water bath to dissolve it completely and make a homogeneous solution. Place slide on a leveled glass plate. Then with a prewarmed fast delivery pipette, pour carefully 2 ml of gel on each slide so that it spreads uniformly. Allow the gel to set. They can be stored in a moist Petri dish and covered with a filter paper.
2. **Loading of samples:** With the help of a piece of coverslip, make a well of about 5 mm deep at about 1 cm away from the margin. Place the slide in the chamber, and connect the ends of the slide with the buffer solution with the help of a filter paper wick. The slide is placed such that the well is at the negative electrode. This is because, proteins are negatively charged in this buffer system and hence they migrate towards positive electrode. The sample to be run is stained usually and 5 μl of the sample is loaded in the well with the help of a capillary tube or micropipette (It is recommended to run a control).
3. **Electrophoretic run:** Care is taken that the electrophoretic chamber is on a leveled surface. The two buffer compartments are filled. The chamber containing the loaded slide(s) is covered with a lid provided and the power is put on. The current is usually adjusted at 7 amp/slide and the voltage is usually maintained at 280 MV. The run is allowed for 45 min–1 hour. Care is taken that the sample does not dribble from the other end.
4. **Fixing & Staining:** After the run, the power is switched off and the slide is taken out. It is put in fixing solution for 15 min and the removed and placed in staining solution for 2 – 3 min. Excess stain is removed by washing and destaining. Scan the slide with a densitometer and quantitate the bands formed.

Clinical Significance

In clinical laboratory, electrophoresis is employed regularly for separation of serum proteins. Albumin has the maximum and γ-globulin has the minimum mobility in the electric field.

Abnormal Patterns in Disease

A number of pathologic conditions are associated with a characteristic serum protein electrophoresis pattern.

1. **Hepatic Cirrhosis:**
   When liver function is sufficiently diminished, protein synthesizing capacity is compromised and concentrations of albumin and proteins in the alpha and beta bands are decreased. An additional common finding is beta-gamma bridging due to increased IgA.
2. **Nephrotic Syndrome:**
   Renal disease involving the glomeruli is always associated with increased urinary protein loss. When protein loss is greater than 3 – 4 g/day, the protein synthesizing capacity of the liver is exceeded and hypoproteinemia, accompanied by anasarca, develops to cause the nephrotic syndrome. The massive urine protein loss is due to increased permeability of glomeruli to protein. The permeability increase may

be minimal so that only albumin and other smaller molecular weight proteins are selectively filtered (selective nephrosis, as in Minimal change disease) or may be greater so that larger proteins are also filtered (nonselective nephrosis, as in membranous glomerulonephritis) as is the case in the example shown. Alpha-2-macroglobulin is sufficiently large so that it is not filtered and increased synthesis (from the general hepatic protein synthesis) causes its accumulation. Lipoproteins are also sufficiently large to accumulate and hyperlipidemia is a characteristic of the nephrotic syndrome, although lipoproteins are not stained with the protein stain used in visualizing proteins.

3. **Alpha-1-Antitrypsin Deficiency:**
   A genetic defect causes a deficiency of alpha-1-antitrypsin. The antiprotease deficiency results in a propensity to develop emphysema. Since alpha-1-antitrypsin is the major component of the alpha-1 band, deficiency is suggested by a reduced alpha-1 band. Deficiency is confirmed by specific immunochemical quantification.
4. **Acute Inflammation:**
   The alpha-1 and alpha-2 bands are increased during the inflammatory response from increased hepatic synthesis of acute phase reactant proteins.
5. **Chronic Inflammation:**
   Immunoglobulin synthesis by antigen activated B lymphocytes transformed to plasma cells is demonstrated by the increased polyclonal gamma band.
6. **Immunoglobulin Deficiency:**
   Deficient immunoglobulin synthesis is revealed by a markedly diminished gamma band. Effected individuals are prone to recurrent infection.
7. **Monoclonal Gammopathy (Multiple Myeloma):**
   An unusually sharp band in the gamma region strongly suggests the presence of a homogeneus immunoglobulin and, thus, the malignant proliferation of plasma cells from a single cell (multiple myeloma) in contrast to the broad, heterogeneous, or polyclonal, gamma band as exhibited above in chronic inflammation from immunoglobulin synthesis by many different clones of plasma cells. Homogeneous immunoglobulins are also found in Waldenstrom's macroglobulinemia (where the sharp gamma band is always IgM).

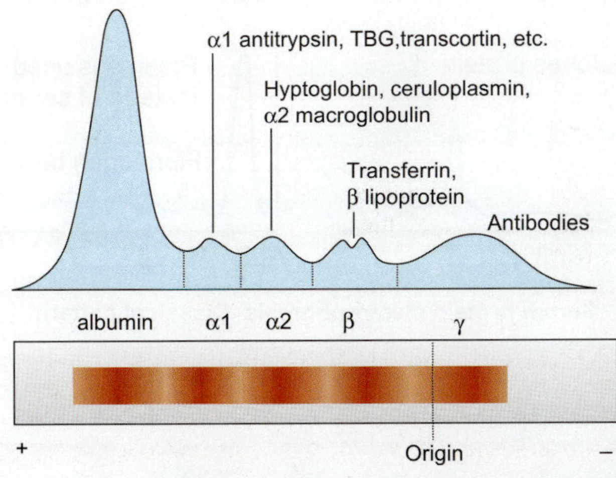

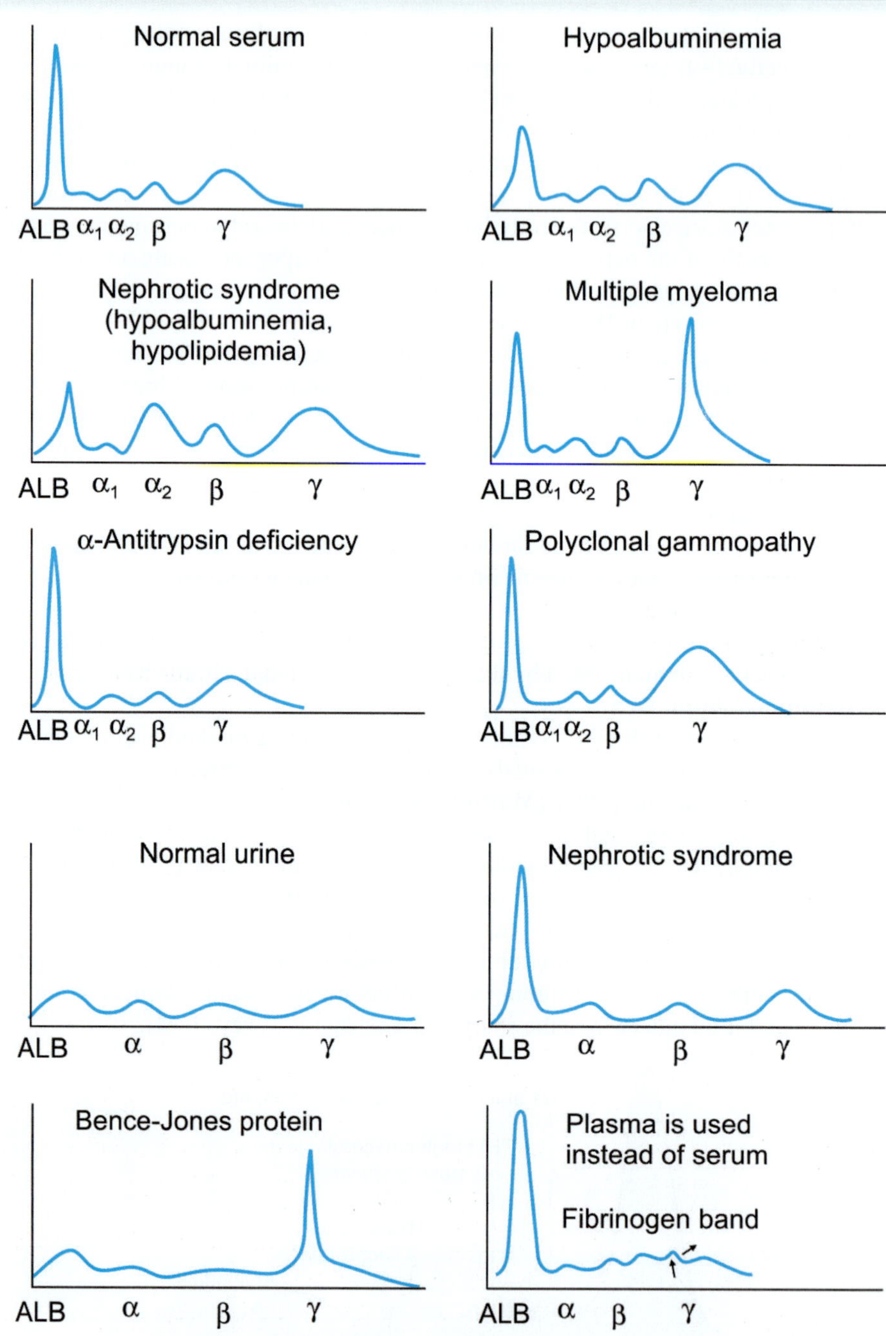

Serum protein electrophoresis–Classical pattern

Chromatography

The word chromatography is derived from Greek words; chroma= color, graphein= to write. This technique was first invented by Michael Tswett in 1906. He separated chlorophyll from other plant pigments using this technique.

Definition and Principle: Chromatography is a technique used to separate different components of a mixture based on their relative distribution between two phases, a stationary phase and a mobile phase. The relative distribution between the two phases is represented mathematically by a term called distribution coefficient or partition coefficient (K_f).

$$K_f = \frac{\text{Concentration of substance in solvent A}}{\text{Concentration of substance in solvent B}}$$

Due to this difference in K_f values of the various components of a mixture, their mobility in two phases differs. A measure of this mobility is called R_f (resolution factor) value.

$$R_f = \frac{\text{Distance travelled by solute}}{\text{Distance travelled by solvent}}$$

A substance which is more distributed in mobile phase will migrate more than the one which is more distributed in stationary phase. This will lead to their separation. Hence, when a mixture reacts with two immiscible phases, its components distribute themselves differentially according to their K_f value leading to their different migrations and R_f value. This can be used to identify and separate the components of a mixture. For instance, different amino acids have different R_f values in a given biphasic system such as of water (stationary phase) and butanol (mobile phase) solvent system. However, to identify the resolved components they are made to react with coloring agents, e.g. ninhydrin for amino acids.

Types of Chromatography: The interaction between stationary phase and mobile phase is often employed in the classification of chromatography.

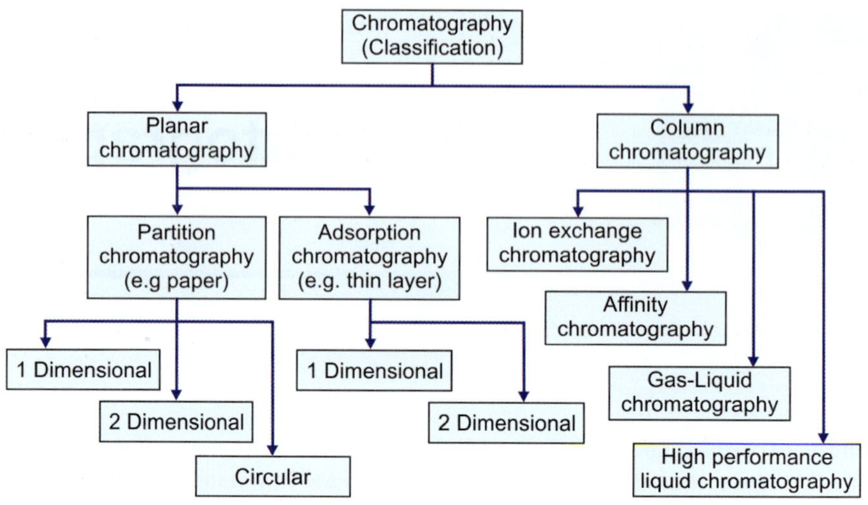

SEPARATION OF AMINO ACIDS BY PAPER CHROMATOGRAPHY

Aim: To separate individual amino acids from a mixture of amino acids by paper chromatography.

Requirements

1. Filter paper (Whatmann No. 1)
2. Pencil
3. Chromatography chamber
4. Amino acid mixture.
5. Solvent: Butanol, acetic acid and water in 4:1:5 ratio
6. Glass capillaries or auto pipette
7. Ninhydrin solution (0.1 %): 100 mg ninhydrin in 100 ml distilled water.

Procedure

1. **Saturation of chromatography chamber:** The solvent is placed in a trough provided in the chamber and the chamber is tightly closed. The chamber is allowed to saturate for 1 hour.
2. **Preparation of chromatogram:**
   a. Cut chromatography paper into rectangular strips and mark a line on the paper at about 2 – 3 cm from the bottom. Marking should always be done with pencil, otherwise the solvent will carry away the markings and there will be unnecessary contamination. Label the paper with its corresponding sample to avoid confusion. Clean your hands and wear gloves before handling chromatography paper.
   b. With the help of capillary tube, take sample and place a spot on the starting line (line drawn in first step). While loading the sample, care should always be taken to avoid spilling.

c. Now, place the chromatography paper in the developing chamber, which contains the solvent or the mobile phase. While placing the paper, it is important that the solvent level doesn't reach the starting line or the sample spots. Make sure that the paper is suspended without touching the sides of the chamber; otherwise it will lead to poor separation. Seal the chamber in a proper manner.

d. The solvent rises up the paper or the stationary phase by capillary action and dissolves the sample. The components of the sample move along with the solvent in upward direction. The speed of movement depends on two factors, the attraction of the solvent molecules to the paper and the differential absorption of the solute components in the solvent. The more attraction or affinity a component has, the slower it moves up and vice-versa. Thus, different components cover different distances within the same time.

e. Check if the solvent has reached near the top level of chromatography paper. Remove the paper when it reaches the top and mark the level with pencil. This level or height is called the "solvent front". Examine the different spots of varied colors. Each spot represents a specific component of the sample. Sometimes, spots are not distinct to locate. In such condition, the paper should be viewed using UV light, ninhydrin or iodine vapors. Carefully circle the spots with pencil.

3. **Resolution factor (R_f) Calculation:** Resolution factor (R_f) is the ratio of the distance travelled by the substance to the distance travelled by the solvent. R_f value is always between 0 and 1 and has no unit. The R_f value of unknown compounds is compared with the R_f value table of known compounds for identification. Measure the distances of the solvent front and also the distances travelled by the components (take the center point of the spots). Calculate the retention factors of the components by using the relation, R_f = distance travelled by the substance/distance travelled by the solvent. Since the distance covered by the components varies, the resulted R_f values will also vary. Compare and match the R_f values of the unknown components with R_f value table and identify the substances present in the particular sample.

APPLICATION OF CHROMATOGRAPHY

- Separation of proteins
- Separation of carbohydrates
- Separation of nucleic acids
- Separation of fats
- Separation of vitamins
- Separation of drugs.

ADVANTAGES AND DISADVANTAGES

As we have seen, paper chromatography is easy to carry out. Paper chromatography can be an ascending chromatography or a descending chromatography, based on the direction of running a chromatogram. A two dimensional chromatography is sometimes performed to resolve components which have near R_f values. Time required for running a chromatogram varies from one hour to several hours. In general, it is used in diagnosis (e.g. amino acidurias), clinical research (e.g. separation of drugs), manufacturing industries and forensic science studies. The disadvantage of paper chromatography is that many a time complex mixtures can't be separated by using this method.

Different forms of chromatography

Mobile Phase	***Stationary Phase***
GAS	LIQUID Gas-Liquid chromatography (GLC)
Gas chromatography (GC)	SOLID Gas-Solid chromatography (GSC)
LIQUID	LIQUID Liquid-Liquid chromatography (LLC)
Liquid chromatography (LC)	SOLID Liquid-Solid chromatography (LSC)

DIFFERENT FORMS OF CHROMATOGRAPHY

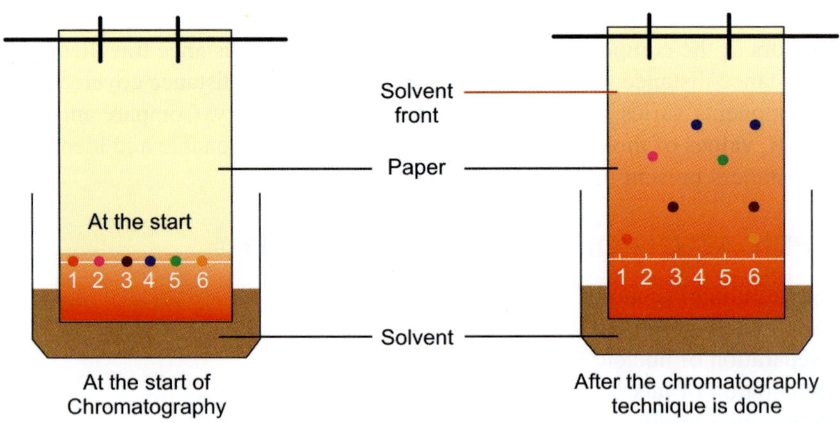

Appendices

I. REFERENCE VALUES OF COMMON ANALYTES

P = Plasma; B = Blood; S = Serum; CSF = Cerebrospinal fluid; pg = picogram; μg=microgram; mg = milligram; g = gram; d = day u = units per liter

Analyte	*Sample*	*Old units*
Ammonia	P/S	< 50 μg/dl
Acid phosphatase (AC), Total	P/S	0.3 – 1 mg/dl
Alanine aminotransferase (ALT/SGPT)	S	10 – 40 U/L
Albumin	S	3.5 – 5 g/dl
Albumin	CSF	10 – 30 mg/dl
Alkaline phosphatase (ALP)	S	40 – 125 U/L
Amino acids, Total	P/S	30 – 50 mg
Amylase	S	35 – 125 U/L
Aspartate amino transferase (AST/SGOT)	S	15 – 40 U/L
Bicarbonate (HCO_3^-)	S	22 – 26 mEq/L
Bilirubin, Total	S	0.2 – 1 mg/dl
Calcium	S	9 – 11 mg/dl
Ceruloplasmin	S	25 – 50 mg/dl
Chloride	S/P	96 – 106 mEq/L
Chloride	CSF	120 – 130 mEq/L
	U	120 – 250 mEq/day
Cholesterol	S/P	150 – 200 mg/dl
Cholesterol (HDL fraction)	S	
	Male	30 – 60 mg/dl
	Female	35 – 75 mg/dl
Cholesterol		
(LDL fraction)	S < 180 mg/dl	

Contd...

Contd...

Analyte	*Sample*	*Old units*
	20 – 29 yrs	60 – 150 mg/dl
	30 – 39 yrs	80 – 175 mg/dl
	40 – 60 yrs	90 – 200 mg/dl
Copper	P	70 – 150 µg/dl
Creatinine	S	0.7 – 104 mg/dl
	U	15 – 25 mg/kg/day
Globulins	S	205 – 305 g/dl
Glucose (fasting)	P	75 – 110 mg/dl
	B	65 – 100 mg/dl
	CSF	50 – 70 mg/dl
Hemoglobin	B	
	Male	14 – 16 g/dl
	Female	13 – 15 g/dl
Iron	S	
	Female:	50 – 130 µg/dl
	Male:	60 – 150 µg/dl
Lipase	S	
pH	B	7.4
pO_2 arterial	B	90 – 100 mmHg
Potassium	S	3.5 -5 mEq/L
Proteins – total	S	6 – 8 g/dl
	CSF	10-30 mg/dl
Sodium	S	136 – 145 mEq/L
Triglycerides, fasting	S	
	Male	50 – 200 mg/dl
	Female	40 – 150 mg/dl
Urea	S	20 – 40 mg/dl
Urea nitrogen	S/P	8 – 20 mg/dl
Uric acid	S/P	
	Male	3.5 – 7 mg/dl
	Female	3.0 – 6 mg/dl
	Children	2.0 – 5.5 mg/ml

II. PREPARATION OF REAGENTS

1. **Molisch's test:**
   i. Molisch's reagent: 1 gm α-Naphthol in 100 ml of alcohol. (1% α-Naphthol in alcohol).
   ii. Conc. Sulfuric acid
2. **Benedict's test:**
   i. Benedict's qualitative reagent: Dissolve 17.3 gms of $CuSO_4$ in 100 ml of distilled water. Dissolve 173 gms of sodium citrate and 100 gms of anhydrous sodium carbonate in 700 ml of distilled water. Add $CuSO_4$ solution slowly stirring. Make up to a liter.
3. **Fehling's test:**
   i. Fehling's solution A: Contains copper sulfate solution. Prepared by dissolving 70 gms of $CuSO_4 . 5H_2O$ in 1 liter of distil water.
   ii. Fehling's solution B: Contains potassium hydroxide and sodium potassium tartrate. Prepared by dissolving 345 gms of sodium potassium tartrate and 130 gms of sodium hydroxide (or 200 gms of potassium hydroxide). Prepare fresh by mixing equal volumes of Fehling's solution A and B.
4. **Barfoed's test:**
   Barfoed's reagent: Boil 900 ml distil water. Add 48 gms copper acetate (with a little water). Mix and add 2 ml of glacial acetic acid. Make up to 1 liter.
5. **Reduction of methylene:**
   i. 1% Methylene blue: 1 gm methylene blue in 100 ml of distil water.
   ii. 5% sodium hydroxide: 5 gms sodium hydroxide in 100 ml of distil water.
6. **Foulger's test:**
   i. Foulger's reagent: 400 gms of urea is dissolved in 80 ml of 40% (w/v) H_2SO_4. 2 gms of stannous chloride is then added and the mixture boiled till clear. After cooling make upto 100 ml with 40% H_2SO_4.
7. **Seliwanoff's test**
   i. Seliwanoff's reagent: 0.05 gms resorcinol dissolve in 33 ml conc. HCl and 67 ml distil water.
8. **Rapid Furfural test:**
   i. 1% alcoholic alpha Naphthol
   ii. Conc. HCl
9. **Bial's test (Tollen's Orcinol Test):**
   Reagents required:
   i. Dilute solution of orcinol in 30% HCl
10. **Osazone test:**
    i. Phenyl hydrazine powder
    ii. Glacial acetic acid
    iii. Sodium acetate powder
11. **Acid Hydrolysis test:**
    i. Conc. HCl

ii. 20% sodium carbonate: 20 gms of sodium carbonate in 100 ml of distil water.

iii. Red litmus paper

Carbohydrate solutions: 1% glucose, 1% fructose, 2% lactose, 2% maltose, 2% sucrose

12. Mercuric Nitrate Precipitation test:

i. 5% mercuric nitrate: 5 gms of Mercury nitrate in 100 ml of distil water

13. Zinc Sulfate Precipitation test:

i. 2% Zinc sulfate: 2 gms zinc sulfate in 100 ml of distil water

14. Precipitation by Esbach's reagent:

i. Esbach's reagent: Dissolve 1 gm of picric acid and 1 gm of citric acid in 100 ml of distill water

15. Precipitation by Sulphosalicylic acid:

i. 20% sulphosalicylic acid: 20 gms of sulphosalicylic acid in 100 ml of distil water

16. Precipitation by Ethanol:

i. Ethanol

17. Half-Saturation test:

i. Saturated ammonium sulfate solution

ii. 40% sodium hydroxide: 40 gms of NaOH in 100 ml of distil water

iii. 1% copper sulfate solution: 1 gm copper sulfate in 100 ml distil water

18. Full saturation test:

i. Solid ammonium sulfate

ii. 40% sodium hydroxide

iii. 1% copper sulfate solution

19. Isoelectric precipitation:

i. Bromocresol green indicator: Dissolve 100 mg of Bromocresol indicator powder in 100 ml of alcohol. Add 100 ml of water and one drop of 10% sodium hydroxide and mix

ii. 1% acetic acid: 1 ml of acetic acid in 100 ml of distil water

20. Heller's test:

i. Conc. HNO_3

21. Biuret test:

i. 5% sodium hydroxide: 5 gm of sodium hydroxide in 100 ml of distil water

ii. 1% copper sulfate: 1 gm of copper sulfate in 100 ml of distil water

22. Ninhydrin test:

i. 0.2% Ninhydrin: 0.2 gms of Ninhydrin dissolved in 100 ml of distil water. (This is not stable for more than 2 days).

23. Xanthoproteic test:

i. Conc. nitric acid

ii. 40% Sodium hydroxide: 40 gms of NaOH in 100 ml of distil water

24. Millon's test:

i. 10% Mercuric sulfate in 10% Sulfuric acid: 10 gms of mercuric sulfate in 10 ml of sulfuric acid and 90 ml of distil water.

ii. 1% Sodium nitrite: 1 gm sodium nitrite in 100 ml distil water.

25. Aldehyde test:

i. 1 in 500 commercial formalin: 1 ml of formalin in 500 ml of distil water
ii. 10% Mercuric sulfate in 10% sulfuric acid: 10 gm of mercuric sulfate in 10 ml of sulfuric acid and 90 ml of distil water
ii. Conc. Sulfuric acid

26. Pauly's test:

i. 1% solution of sulfanilic acid in 1M HCl.
ii. 5% Nitrous acid
iii. 1% Na_2CO_3

27. Sakaguchi test:

i. 40% sodium hydroxide: 40 gms of NaOH in 100 ml of distil water
ii. 1% alpha Naphthol: 1 gm of alpha Naphthol in 100 ml of ethanol
iii. Bromine water: 2 ml bromine (highly corrosive) in a liter of distil water

28. Lead Sulfide test:

i. 40% sodium hydroxide: 40 gm of NaOH in 100 ml of distil water
ii. 2% lead acetate solution: 2 gm of lead acetate in 100 ml of distil water

Egg albumin solution: 1 in 6 times egg albumin diluted with water with a pinch of sodium chloride

29. Heat coagulation test:

i. Chlorophenol red indicator: Dissolve 100 mg of chlorophenol red in 100 ml of alcohol. Add 100 ml of distil water and one drop of 10% sodium hydroxide and mix
ii. 1% Acetic acid: 1 ml of glacial acetic acid in 100 ml of distil water
iii. 1% sodium carbonate: 1 gm of sodium carbonate in 100 ml of distil water

30. Precipitation by phosphotungstic acid:

i. Phosphotungstic acid: Dissolve 100 gms of sodium tungstate in 600 ml of distil water. Add 80 ml of orthophosphoric acid. Boil under reflux for one hour. Make up to one liter after cooling.
ii. 40% sodium hydroxide
iii. 1% copper sulfate

31. Test for inorganic phosphorus:

i. 40% sodium hydroxide: 40 gms of sodium hydroxide in 100 ml of distil water
ii. Conc. nitric acid
iii. Solid ammonium molybdate

Gelatin solution: 1% Gelatin solution in water
Casein solution: 1% casein solution with 3 – 4 drops of 5% NaOH
Peptone solution: 1% peptone solution by heating on low flame

32. Urease test:

i. Phenol red indicator: Dissolve 100 mg of phenol red powder in 100 ml of alcohol. Add 100 ml of distil water and one drop of 10% sodium hydroxide and mix
ii. 1% acetic acid
iii. 1% sodium carbonate
iv. Horse gram powder in distil water

33. Hypobromite test:

i. Sodium hypobromite solution: Dissolve 10 gm of NaOH in 990 ml of distil water. Add 10 ml of bromine (Bromine is highly corrosive) and mix

34. Phosphotungstic acid reduction test:

i. Phosphotungstic acid reagent: Dissolve 100 gm of sodium tungstate in 600 ml of distil water. Add 80 ml of orthophosphoric acid. Boil under reflux for one hour. Cool and make upto one liter.

ii. 20% sodium carbonate: 20 gms of sodium carbonate in 100 ml of distil water

35. Jaffe's test:

i. Saturated picric acid solution

ii. 10% sodium hydroxide: 10 gms of sodium hydroxide in 100 ml of distil water

36. Test for Ammonia:

i. 2% sodium carbonate: 2 gms of sodium carbonate in 100 ml of distil water

ii. Red litmus paper

37. Test for chlorides:

i. Conc. Nitric acid

ii. 3% silver nitrate solution: 3 gms of silver nitrate in 100 ml of distil water

38. Test for inorganic phosphorus:

i. Conc. Nitric acid

ii. Solid ammonium molybdate

39. Test for calcium:

i. 1% acetic acid: 1 ml of glacial acetic acid in 100 ml of distil water

ii. 2% potassium oxalate: 2 gms of potassium oxalate in 100 ml of distil water

40. Test for sulfates:

i. Conc hydrochloric acid

ii. 10% barium chloride: 10 gms of barium chloride in 100 ml of distil water

41. Benedict's reduction test:

i. Benedict's reagent: for preparation, refer Expt. No. 2

42. Rothera's test:

i. Solid ammonium sulfate

ii. Solid sodium nitroprusside

iii. Ammonia solution

43. Gerhardt's test (Acetoacetic acid)

44. Benzidine test:

i. Benzidine powder

ii. Glacial acetic acid

iii. Hydrogen peroxide

45. Guaiac test:

i. Hydrogen peroxide 30%

ii. Gum guaiac reagent

iii. Gum guaiac reagent

Prepare fresh before use and use the reagent on the same day:

Gum guaiac ---------- 0.1 g
Ethanol 95% v/v --------- 6.0 ml

Weigh the chemical and transfer it to a bottle of 10 ml capacity. Add the ethanol, stopper the bottle and mix until the chemical is fully dissolved. Caution: the ethanol reagent is flammable; therefore handle it well away from any open flame.

46 **Sulphosalicylic acid test:**
   i. 20% sulphosalicylic acid: 20 gms of sulphosalicylic acid in 100 ml of distil water

47. **Heat coagulation test:**
   i. 1% Acetic acid: 1 ml glacial acetic acid in 100 ml of distil water

48. **Hay's test:**
   i. Sulfur powder

49. **Fouchet's test:**
   i. Fouchet's reagent: Dissolve 25 gms of trichloroacetic acid and 1 gm of ferric chloride in 150 ml of distil water
   ii. 10% barium chloride: 10 gms barium chloride in 100 ml of distil water

50. **Ehrlich's test:**
   i. Ehrlich's reagent: Dissolve 0.7 gm of para dimethyl amino benzaldehyde in 150 ml of conc. hydrochloric acid and 100 ml of distil water (store in brown bottle)
   ii. Sodium acetate powder

51. **Blood glucose estimation:**
   Reagents required:
   i. Ortho-toluidine reagent
      - Glacial acidic acid – 940 ml
      - Ortho-toluidine – 60 ml
      - Thiourea – 1.5 gm (for stability)

      Stable for six months at 2 – 8°C
   ii. Glucose standard: 100 mg/dl in saturated benzoic acid. Stable for one year at room temperature

52. **Estimation of blood urea:**
   Reagents required:
   i. Sodium tungstate 10%
   ii. Sulfuric acid 2/3N
   iii. Diacetyl monoxime 2% solution in 2% acetic acid. Add 2 gms of the solid to about 60 ml of water, add 2 ml of glacial acetic acid, shake to dissolve with slight warming if necessary, and make up with water to 100 ml.
   iv. Sulfuric acid-phosphoric acid reagent. Add 150 ml of 85% phosphoric acid to 140 ml water, mix well and add 50 ml of concentrated sulfuric acid slowly whilst mixing.
   v. Stock standard solution of urea 250 mg/100 ml
   vi. Working standards. Dilute the stock standard 1 to 100 to give solution containing 0.025 mg urea per ml.

53. **Estimation of urinary creatinine:**
   Reagents required:
   i. Picric acid (0.04 m solution) 9.16 grams per liter
   ii. Sodium hydroxide (0.75 N solution)

      Dissolve 30 gms of sodium hydroxide in 1 liter of distil water

iii. Stock solution of creatinine containing 1 mg of creatinine per ml. Dissolve 100 mg of pure dry creatinine in 0.1N hydrochloric acid and make up to 100 ml with the acid. The solution keeps almost indefinitely.

iv. Working creatinine standard for use:
Prepare from the above by diluting 1 ml of stock solution to 100 ml with water.
This contains 0.01 mg of creatinine per ml.

54. Total serum protein estimation:

Reagents required:

i. Stock biuret reagent: Dissolve 45 gms of Rochelle salt in about 400 ml of 0.2N sodium hydroxide and add 15 gms of copper sulfate (small crystals of $CuSO_{4,}.5\,H_2O$) stirring continuously until solution is complete. Add 5 gms of potassium iodide and make up to a liter with 0.2N sodium hydroxide.

ii. Biuret solution for use: Dilute 200 ml of stock reagent to a liter with 0.2N sodium hydroxide which contains 5 gms of potassium iodide per liter.

iii. Tartrate iodide solution: Dissolve 9 gms of Rochelle salt in 0.2N sodium hydroxide containing 5 gms of potassium iodide per liter.

iv. Bovine or human albumin standard, 2 mg/ml.

55. Cerebrospinal fluid:

a. Determination of CSF total proteins:

Reagents required

i. Sulphosalicylic acid: 3 gm/dl sulphosalicylic acid.

ii. Normal saline: 0.85 gm/dl of sodium chloride

iii. Stock protein standard: 6 gms/dl of bovine albumin stable for one year when refrigerated.

iv. Working protein standard (60 mg/dl): Prepare by mixing 0.1 ml of the stock standard and 9.9 ml of normal saline. This should be prepared fresh.

b. Qualitative tests for globulins:

Reagents required:

i. Saturated ammonium sulfate solution. Dissolve 85 gm of ammonium sulfate in 100 ml of hot distilled water, allow to cool overnight and filter. Keep in a well stoppered bottle.

ii. Pandy's solution: Prepare by dissolving 10 gms of phenol in 150 ml of distilled water. The solution should be clear and colorless.

c. Determination of CSF chloride:

Reagents required:

i. Silver nitrate solution: 30 mEq/L
Dissolve 5.10 gms of Silver nitrate in 1 liter of distil water.

ii. Potassium chromate: 10% solution
Dissolve 10 gms of Potassium chromate in 100 ml of distilled water

Artificial CSF preparation: Add 30 mg of Albumin, 60 mg of glucose and 730 mg of NaCl to 100 ml of distilled water and mix properly.

Index

Page numbers followed by *f* refer to figure